AN INQUIRY INTO

The Human Prospect

Books by Robert L. Heilbroner

Marxism: For and Against
Beyond Boom and Crash
An Inquiry into the Human Prospect
Business Civilization in Decline
Between Capitalism and Socialism
The Economic Problem
The Limits of American Capitalism
Understanding Macroeconomics
A Primer on Government Spending
(with Peter L. Bernstein)
The Great Ascent
The Making of Economic Society
The Future as History
The Quest for Wealth
The Worldly Philosophers

Robert L. Heilbroner

AN INQUIRY INTO

The Human Prospect

Updated and Reconsidered for the 1980s

W · W · NORTON & COMPANY

NEW YORK · LONDON

Library of Congress Cataloging in Publication Data
Heilbroner, Robert L
    An inquiry into the human prospect.
    Includes index.
    1. Civilization, Modern—1950- 2. Regression
(Civilization) I. Title.
CB428.H44   1974   909.82   79-25610
ISBN 0-393-01371-5 CLOTH
ISBN 0-393-95139-1 PAPER

1 2 3 4 5 6 7 8 9 0

Contents

	Foreword	7
ONE	Initial Reflections on the Human Prospect	11
	AFTERWORD	25
TWO	The External Challenges	31
	AFTERWORD	58
THREE	Socio-Economic Capabilities for Response	77
	AFTERWORD	111
FOUR	The Political Dimension and "Human Nature"	119
	AFTERWORD	144
FIVE	Final Reflections on the Human Prospect	149
	AFTERWORD	166
	Postscript: What Has Posterity Ever Done for Me?	179
	Acknowledgments	187
	Index	189

Foreword

An Inquiry into the Human Prospect was written between July 1972 and August 1973, a period scarcely ten years away that now seems light-years distant. Perhaps it can best be described as a time when a great national illusion was gradually destroyed—the illusion that an invisible shield surrounded the United States, like those force fields in science fiction. This shield, we believed, held at bay the brutalities and irrationalities that seemed to be part of the life of other nations, but not our own. It bestowed on American life an invulnerability against the ravages that history worked elsewhere in the world. It produced in us the mixture of innocence and assurance with which the best Americans faced the future, as well as the noisy arrogance with which the worst did.

During the period when this book was conceived, that shield gradually disappeared—or rather, our belief in it evaporated. From our terrible experiences at home as well as abroad, from a smugness and mendacity in high government as demoralizing in their own way as the atrocities in Vietnam, came the slow realization that there was nothing outrageous or brutal in

Africa or Asia or Latin America that could not also be found in the United States. During those years, the recognition dawned that the American way of life was no more exempt from erosion or obliteration than those of other nations; indeed, that by virtue of our prominence and wealth, our innocence and self-assurance, we were, in a sense, singled out for exposure to history's challenges.

I begin by harking back to the mood of puzzlement and despair that formed so important a part of the climate of the early 1970s because it had much to do with the reading, as well as the writing, of the *Inquiry*. Of course my readers questioned many things in its pages. Yet, there was one thing that no one called to account. This was the validity of the wager I made on the first page of the book—a wager that a reader picking up my tract and facing its terrible opening question would not fling the book away in disbelief. The question asked: Is there hope for man? Whatever the reactions to the way in which I addressed myself to it, no one thought the query itself merely rhetorical or foolish.

Today, nearly a decade after that period of awakening, I speak to an audience that shares a distinctly different mood. Some of the challenges that were startling and disconcerting when the *Inquiry* first appeared have by now achieved a certain dullness of familiarity: it is no longer news that we face an energy crisis— indeed, the recognition is gradually dawning that it is not a "crisis" from which, like a sick patient, we shall recover in due time, but a new condition, a permanent change in our capacities. Other problems have receded temporarily in the background: for instance, we have become inured to the spread of nuclear weaponry to minor countries, a matter of headline significance when the *Inquiry* was being written and India, to worldwide consternation, detonated an atom bomb.

Most important of all, the attitudes of innocence and assurance that were then under such painful attack have by now achieved an almost antiquarian status. Is it possible that only ten years ago most Americans fervently believed that their government would never tell a lie? Is it credible that we used to talk seriously about America's unchallengeable global power? This is some measure of the distance we have come, or of the depths to which we have fallen, or of the progress we have made, depending on how you look at it.

At any rate, the change in our perceptions has influenced the manner in which I have approached the task of "updating" *The Human Prospect.* As its new subtitle indicates, I have tried not merely to make it more current, but to reconsider it—that is, to reflect back on the plausibility of the original arguments from the altered vantage point of the impending 1980s. For this reason I have decided to leave the original text alone, trifling changes aside. Instead, following each chapter I have added an Afterword, where I can scrutinize the facts and the train of thought, not only to make corrections where these are needed, but to try to bridge the gap between the mood I shared with my original readers and the different mood of today.

My book has had the good fortune to remain "contemporary" for almost a decade. I hope that this new inquiry into the *Inquiry* will extend that contemporaneity for a little while longer, events permitting.

ROBERT L. HEILBRONER

October 1979, New York City

ONE

Initial Reflections on the Human Prospect

THERE IS A question in the air, more sensed than seen, like the invisible approach of a distant storm, a question that I would hesitate to ask aloud did I not believe it existed unvoiced in the minds of many: "Is there hope for man?"

In another era such a question might have raised thoughts of man's ultimate salvation or damnation. But today the brooding doubts that it arouses have to do with life on earth, now, and in the relatively few generations that constitute the limit of our capacity to imagine the future. For the question asks whether we can imagine that future other than as a continuation of the darkness, cruelty, and disorder of the past; worse, whether we do not foresee in the human prospect a deterioration of things, even an impending catastrophe of fearful dimensions.

That such a question is in the air, hovering in the background of our minds, is a proposition that I shall not defend by citing bits of evidence from books, articles, and the like. I will rest my case on the reader's own response, gambling that my initial assertion does not generate in him or her the incredulity I should feel were I to open a book whose first statement was that the prevailing mood of our times was one of widely shared optimism. Thus I shall simply start by assuming that the reader shares with me an awareness of an oppressive anticipation of the future. The nature of the evidence on which this state of mind ultimately rests will be the subject of our next chapter. But the state of mind itself must be looked into before we can proceed to an examination of the evidence, for our initial perspective enters into and colors the assessment we make of the "objective" data. Let us therefore open our inquiry into the human prospect by taking stock of our current anxiety.

Evidences of this anxiety are to be found in many advanced nations. Here, however, I shall focus on the American mood, not only because I know it best but because it has been itself a bellwether of feelings elsewhere. If we now look into the American state of mind, I think we can find three main sources, or perhaps three levels of explanation, for the pall that has fallen over our spirits.

The first of these I will call *topical,* to refer to a barrage of confidence-shaking events that have filled us with a sense of unease and foreboding during the past decade or so. No doubt foremost among these events has been the experience of the Vietnam war, an experience that has undermined every aspect of American life—our belief in our invincible power, our trust in our government, our estimate of our private level of morality. But the Vietnam war was only one among many such confidence-shaking events. The explosion of violence in street crime, race riots, bombings, bizarre airplane hijackings, shocking assassinations has made a mockery of the television image of middle-class American gentility and brought home with terrible impact the recognition of a barbarism hidden behind the superficial amenities of life.

Perhaps even more important among these topical causes for our pessimistic frame of mind has been yet another development of the recent past—the failure of the present middle-aged generation to pass its values along to its children. The ubiquitous use of drugs, the extreme sexual relaxation, the defiantly unconventional modes of dress, the unprecedented phenomenon of "dropping out," especially among the children of the most successful classes, all have added their freight of disquiet and disconcert to the mood of our times.

When I call these causes of our present mood "topical," I do not imply that they are mere surface phenomena. Some of these manifestations may be no more than those curious societal epidemics that have often raged and then burned themselves out; others seem to have deeper roots and to signify changes of longer duration. By topicality I refer, rather, to the fact that these events have directly entered our daily lives, whether through newspaper headlines or by personal experience. They are events that have become part of our day-to-day existences, the conversational fare of a million breakfast tables, instilling in us a feeling of dismay, often bordering on despair.

I do not think, however, that we can account for our present mood solely in terms of these topical blows. Hence I call to our attention a second source of our present mood. This is a series of *attitudinal* changes that underlie and reinforce the topical events—changes that have not presented themselves as immediate existential concerns but have made themselves felt nonetheless as part of our inarticulated consciousness.

Somewhat arbitrarily, I select two of these attitudinal changes as being of central importance. The first is a loss of assurance with respect to the course of social events. The present generation of adults passed its formative years in a climate of extraordinary self-confidence regarding the direction of social change. For the oldest among us, this security was founded on the lingering belief in "progress" inherited from the late Victorian era—a belief suffused for some with expectations of religious or moral perfectibility, for others with more cautious but no less sustaining beliefs in the solid prospects for bourgeois society.

For the middle-aged, educated, as I was, in the 1930s, this Victorian heritage was already regarded as a period piece, battered first by World War I, then

14

dealt its death blow by the Great Depression. But its comforting assurance had been replaced by an equally fortifying belief. This was the view that history, working like a vast organic machine, would produce a good socialist society out of a bad capitalist one. And for the younger adults, who formed their ideas in the 1940s and 1950s when this Marxian vista was itself regarded as an antique, reassurance was still provided by a pragmatic, managerial approach to social change. This was a time when one spoke of social problems as so many exercises in applied rationality: when economists seriously discussed the "fine tuning" of the economy; when the repair of the misery of a billion human beings was expected to be attained in a Decade of Development with the aid of a few billion dollars of foreign assistance, some technical advice, and a corps of youthful volunteers; when "growth" seemed to offer a setting in which many formerly recalcitrant problems were expected to lose their capacity for social mischief.

Today that sense of assurance and control has vanished, or is vanishing rapidly. We have become aware that rationality has its limits with regard to the engineering of social change, and that these limits are much narrower than we had thought; that many economic and social problems lie outside the scope of our accustomed instrumentalities of social change; that growth does not bring about certain desired ends or arrest certain undesired trends. One of these unmanageable events is the apparently unstoppable inflation that we witness in every industrialized capitalist nation. Another is the seemingly uncontrollable force of racial hatred, evident not only at home but in the relations of Hindus and Moslems, Jews and Arabs, Africans and Africans. Yet another is the stubborn resistance of world poverty to the ministrations of foreign aid, a phenomenon that we may perhaps understand better when we reflect on our inability to prevent the decline of some American cities into wastelands.

Hence, in place of the brave talk of the Kennedy generation of managerialists—not to mention the prophets of progress or of a benign dialectical logic of events—there is now a recrudescence of an intellectual conservatism that looks askance at the possibilities for large-scale social engineering, stressing the innumerable cases—for example, the institutionalization of poverty through the welfare system, or the exacerbation of racial friction through efforts to promote racial equality—in which the consequences of well-intentioned acts have only given rise to other, sometimes more formidable problems than those they had set out to cure.

Yet I do not believe that this second source of the erosion of confidence would by itself account for the pall that hangs over us were it not combined with another attitudinal change. This is our startled awareness that the quality of our surroundings, of "life," is deteriorating. Of all the changes in our background awareness, perhaps none is so important as this.

One aspect of this new awareness is a fear that we will be unable to sustain the trend of economic growth very much longer. The advent of an energy "crisis" alerts us to the prospect of a ceiling on industrial production, imposed by an inability to overcome the rapidly diminishing returns of a natural world that is being mined more voraciously each year. Such a prospect brings the troubling consideration of how we would manage the direction of events if economic growth—the central pillar of support for the sanguine views of Victorians, traditional Marxists, and managerialists alike—were forced to come to an early end.

But this prospect, though it may be the more immediate cause of our new-found concern with growth, is fundamentally less troubling than another recently recognized state of affairs. This is the stunning discovery that economic growth carries previously unsuspected side effects whose cumulative impact may be more deleterious than the undoubted benefits that growth also brings. In the last few years we have become apprised of these side effects in a visible decline in the quality of the air and water, in a series of man-made disasters of ecological imbalance, in a mounting general alarm as to the environmental collapse that unrestricted growth could inflict. Thus, even more disturbing than the possibility of a serious deteri-

17

oration in the quality of life if growth comes to an end is the awareness of a possibly disastrous decline in the material conditions of existence if growth does not come to an end.

Perhaps the combination of these topical and attitudinal elements is enough to account for the dark mood of our time. But I shall nevertheless advance a third reason, although I suspect it only flickers, so to speak, in our consciousness. It is a *civilizational* malaise that enters into our current frame of mind.

For some time, observers skeptical of the panacea of growth have wondered why their contemporaries, who were three or five or ten times richer than their grandparents, or great-grandparents, or Pilgrim forebears, did not seem to be three or five or ten times happier or more content or more richly developed as individual human beings. This skepticism, formerly the preserve of a few "philosophically minded" critics, has now begun, I believe, to enter the consciousness of large numbers of men and women.

The skepticism had a certain ring of hypocrisy so long as the great majority of men lived in a condition of low material attainment and static expectations. Only in the last century or so, as great masses of people have moved "up" the scale—each generation consuming food in quantity and quality superior to that of the classes above them in the preceding generation, each generation clothed in materials whose variety, color, fineness, and abundance surpassed the garb of all but the very wealthiest figures of its youth, each generation able to enjoy a degree of mastery over death that would have appeared miraculous to its progenitors, each generation able to move about the surface of the earth or to command the powers of nature in ways that would have struck the previous generation with awe—only then could the warnings of the philosophers as to the ultimate inadequacy of material possessions be tested in reality and, after an initial period of disbelief, discovered to be true.

The civilizational malaise, in a word, reflects the inability of a civilization directed to material improvement—higher incomes, better diets, miracles of medicine, triumphs of applied physics and chemistry—to satisfy the human spirit. To say as much is not to denigrate its achievements, which have been colossal, but to bring to the forefront of our consciousness a fact that I think must be reckoned with in searching the mood of our times. It is that the values of an industrial civilization, which for two centuries have given us a sense of *élan* and purpose, now seem to be losing their self-evident justification. To date, the doubts and disillusions as to that civilization are only faint breezes that stir the leaves of the tree and will certainly not uproot a way of life anchored deeply in the earth of our beings. But the breezes blow and the stirrings they cause must be added to the sense of sometimes indefinable unease that is so much a part of our age.

It must be clear from these introductory remarks that I do not pose the question at the outset of this book—"Is there hope for man?"—as a mere rhetorical flourish, a straw figure to be dismantled as we proceed into more "serious" matters. The outlook for man, I believe, is painful, difficult, perhaps desperate, and the hope that can be held out for his future prospect seems to be very slim indeed. Thus, to anticipate the conclusions of our inquiry, the answer to whether we can conceive of the future other than as a continuation of the darkness, cruelty, and disorder of the past seems to me to be no; and to the question of whether worse impends, yes.

But all that remains yet to be demonstrated, or at least spelled out in some detail. And here we encounter a problem that must be faced before we plunge into the task of exposition. How are we to deal with the elements of wish and fear, prejudice and bias, charity and malice that come flooding into an inquiry such as ours, threatening to divert it, despite our best intentions, toward some outcome that we favor from the start?

The problem caused by the intrusion of subjective values into its inquiries has always troubled social science, which has struggled, without too much success, to attain the presumed "value-free" objectivity of the natural sciences. Alas, this ambition fails to take into account that the position of the social investigator differs sharply from that of the observer of the natural world. The latter may stake his reputation as he regards the stars through his telescope or the cells through his microscope, but he is not himself morally embedded in the field he scrutinizes. By contrast, the social investigator is inextricably bound up with the objects of his scrutiny, as a member of a group, a class, a society, a nation, bringing with him feelings of animus or defensiveness to the phenomena he observes. In a word, his position in society—not only his material position but his moral position—is implicated in and often jeopardized by the act of investigation, and it is not surprising, therefore, that behind the great bulk of social science we find arguments that serve to justify the existential position of the social scientist.

These difficulties must certainly affect an inquiry such as this one, not only on the part of myself, whose position, values, and interests shape and influence my perceptions, but equally on the part of the reader, who brings with him like considerations of social identification and similar vulnerabilities of moral posture.

To these difficulties I must add another, no less vexing. It concerns the "facts" toward which we must try to present an impartial and disinterested gaze. Unhappily, these facts are themselves a special problem for our inquiry. I shall try, of course, to base my argument on findings that will withstand the demolition of next year's research. But there is an aspect to the problem that goes beyond the obvious pitfalls in marshaling and weighing the evidence. It arises because much in our estimate of the human prospect must rest on generalizations for which there exist no objective data at all.

For the gravity of the human prospect does not hinge alone, or even principally, on an estimate of the dangers of the knowable external challenges of the future. To a far greater extent it is shaped by our appraisal of our capacity to meet those challenges. It is the flexibility of social classes, the resilience of socio-economic orders, the behavior of nation-states, and ultimately the "nature" of human beings that together form the basis for our expectations, optimistic or pessimistic, with regard to the human outlook. And for these critical elements in the human prospect there are very few empirical findings on which to rest our beliefs. We possess little or no "hard" information about the propensities of nation-states to peace and war, about the stubbornness or adaptability of social classes, or about the malleability of individual beings, except for those frail generalizations that we assemble from our real and vicarious life experience—itself biased, as we have said, by our situation within society and our private predilections. Thus, as regards the most important element of an effort to assess the prospect for man we have no guide but ourselves, and are thrown back, willy-nilly, to criteria that trouble us by virtue of their subjective foundation.

Here, as before, we encounter problems for which there is no solution other than the limited safeguards offered by self-scrutiny and a determined effort to subordinate our private interests to the superior claims of a "dispassionate analysis." I raise them, nonetheless, because I believe that not the least difficult part of an effort to discuss the human prospect is that of disengaging ourselves, either as partisans or as apologists, from the social situation in which we find ourselves, or from the social situation in which we could imagine ourselves in the future. Such considerations of self-interest may, and perhaps should, powerfully influence the point of view we take in advocating or opposing certain kinds of social change, but they can only play a distorting role when we try to stand aside from our private fates and reflect on the probable course of, and causes for, events, whether they are favorable for ourselves or not.

Talleyrand once remarked that only those who had lived in the *ancien régime* could know what "*les douceurs de la vie*" could be. He was referring to the *douceurs* of a court in which elegance and extravagance knew no bounds, and in which the wealthy and highly placed could indulge their whims and caprices with an abandon that we can only look back upon with the mixed feelings with which we regard the indulgence of all infantile desires.

In our period of history, however, it may well be that the threatened *douceurs* are those of an intellectual milieu in which the most extravagant and heretical thoughts can be uttered, if not in perfect safety (what society does not take *some* safeguards against its own destruction?), at least to a degree that has few parallels in history. Now let us suppose that the exigencies of the future, as we shall trace them out, point to the conclusion that only an authoritarian, or possibly only a revolutionary, regime will be capable of mounting the immense task of social reorganization needed to escape catastrophe. Might it not then be argued that the quasi-military devotion and sacrifice of such a task would be vitiated if the masses were exposed to the disagreements and diversions of intellectuals who strayed from, or opposed, the official line? Indeed, might not the people of such a threatened society look upon the "self-indulgence" of unfettered intellectual expression with much the same mixed feelings that we hold with respect to the ways of a vanished aristocracy—a way of life no doubt agreeable to the few who benefited from it, but of no concern, or even of actual disservice, to the vast majority?

I raise this issue not to debate its merits but to bring home as sharply as I can the kinds of defenses, arguments, and rationalizations to which our analysis will lead on more than one occasion. For were the necessary sacrifice not freedom of expression but freedom of acquisition, I imagine that a quite different set of emotions and individuals would be outraged.

Let me therefore forewarn the reader that he must be prepared to face problems in which values and beliefs precious to him may be assaulted by overriding

claims of human survival, and that he must therefore
be prepared seriously to consider painful conclusions
if he is not simply to substitute preference for analysis. Perhaps I should add that many conclusions in
this book have caused great pain to myself, a fact
which in no way vouches for their cogency but does at
least argue that the human prospect, as I have come to
see it, is not one that accords with my own preferences
and interests, as best I know them.

AFTERWORD

I AM ACUTELY AWARE that things do not
look quite the same today as they did
when I wrote this opening chapter. The topical disorders that played so important a role in creating the
atmosphere of siege have faded away. Radical life-
styles have given way to conservative ones. Hair is
short again. There is still violence in America, but for
money, not for ideas or ideals. Parents are less defiantly confronted. More to the point, the political
charge that electrified everything in those days has
largely dissipated, and the prevailing mood is now one
of apathy, or of a marshaling of energies for conservative, not radical, causes.

The reasons for this change in our topical concerns are not difficult to identify. To begin with, some
of the radical efforts of the late sixties made their

mark: women and blacks are not yet equal citizens, but they are more equal. Street language is now parlor language; our views on sex have changed; the nation is not as "up-tight" as it was. Whether this is all gain is hard to say, but it has lowered the tension of that period of fierce transition in which every new thrust of a frustrated generation seemed a threat aimed at the heart of society.

Perhaps more important, the momentum of the genuine—not the superficial—radical impulse has slowed, for reasons that we recognize, without wholly understanding them. Pendulums swing; fires burn down. As history has demonstrated so often, nothing is more difficult than to maintain the driving force of change. The enemy of revolution is not just active opposition, for that may fan the flames of revolutionary purpose; it is inertia, indifference, the diversions and distractions of daily life. Oscar Wilde once remarked that the trouble with socialism was that it took too many evenings, and there is a bitter truth in this cynicism.

In retrospect, it is surprising that the political drive of the 1960s—so intense, so limitless, so threatening when this book was being written—lasted as long as it did. It was never a true political movement: it had no class basis, not even an organizational structure. Thus when initial enthusiasms faded, there was no steady energy of self-interest, no systematic program, not even a well-defined group of leaders to keep the movement alive. The widespread fear that radicalism would displace our conventional values or institutions was only a sign of our insecurity, not a reaction to an actual threat.

A second marked difference between the mood ten years ago and today has to do with our attitudinal stance. Many of the problems are the same. Urban blight continues to resist our efforts to arrest it. The natural environment continues to deteriorate. Government continues to produce inadequate results. What is different is our reaction to these things. When I wrote the initial chapter originally, we were still experiencing our dwindling ability to effect social change as a blow to our self-esteem. In recent years that has been replaced by an intensification of an attitude already visible when I wrote this chapter—a willing acquiesence before this loss of control, even an aggressive assertion that we should never have entertained such ambitions in the first place. Meanwhile, as our confidence in social control fades, our indulgence in self-gratification rises: indeed, one suspects that there is more than a little truckling to the acquisitive instinct in our retreat from a belief in a need to seek the common good.

This general relaxation of mood and posture does not, of course, change the nature of the human prospect. But it may considerably alter our attitude to it. I wonder, for example, whether the rumblings of a civilizational malaise are as widely heard as before. When this book was first published, it was subject to many criticisms, but not to the criticism that no such malaise existed. I am inclined to doubt that an equally accepting reception greets this diagnosis today—not because I believe the deepest ailments of our society have been cured, but because the changes in our attitudes and topical concerns no longer incline our awareness to such somber possibilities.

Our present mood may change. But it suggests that if I am to engage my readers today I must find some new shared experience to serve in the stead of those that spoke to my readers of the early 1970s. There is no question what that common experience is. It is the economic convulsions that we have lived through since this book was first published. During this time, we have seen the stock market, the traditional barometer of economic self-confidence, register the most precipitious short-run decline in its entire history, actually outdoing the Great Crash of 1929 in that of late 1978. Of greater import, we have watched the American dollar, for the first time in its long history, become a currency that foreigners wanted to get rid of, not accumulate. And by far the most anxiety-rousing of all, we have seen inflation change from a mild disorder, sometimes even regarded as a benign affliction, to a worldwide condition that threatens to disrupt economic life to its very roots and that seems to defy every treatment we have devised for it.

I put forward these economic worries not only because they are important in their own right, but because I think they may convey to readers today the same shared, but unspoken unease for which I used other examples and experiences in the original text. For the economic problems have not only left us injured and bruised; they have roused in us the question of whether we were being shaken and tossed about because of deep-lying subterranean slippages, of whether the crisis that is still so much part of our lives does not reach down to foundations below our lives.

Because the economic convulsions of the last years are understandably in the forefront of our minds, I will be giving them special attention in some of the Afterwords to follow. Yet I mention them not to focus attention on them at this point, but to conjure up the necessary frame of mind with which to begin an examination of the larger challenges facing humankind. *For we need an awareness of what it means to live in dangerous times, if we are to take the full measure of the external challenges that frame the human prospect.* The sense of peril that we have gained from our experience with the economic crisis will help us as we now examine those challenges, first as they appeared to me not quite a decade ago, then as they appear today.

TWO

The External Challenges

I HAVE SPOKEN so far of the "mood of our
times" and of the personal considerations
that make an appraisal of the human prospect so
uncommonly difficult. But I have not yet set forth,
except in the most general way, the nature of the
challenges that we perceive in the world about us. As I
have already indicated, the elements of danger in the
human prospect are by no means all located in "external" threats, but in our "internal" capacity for
response to those threats. Yet certainly the roots of
our current anxiety spring in the first instance from
dangers that we discern in the world around us, and it
is therefore to these that our attention must first be
addressed.

If we were asked to identify the principal "external" causes for the mood that assails us, I think that
three aspects of the current human predicament would
be unanimously selected. The first is a problem so well
known that it has almost lost its power to shock,
perhaps because attention has been focused largely on
its humanitarian rather than its political implications. I

refer to the demographic outlook for the next two to three generations.

World population is today roughly 3.6 billion. About 1.1 billion live in areas where demographic growth rates are now tapering off, so that, barring unanticipated reversals in the trends of fertility and mortality, we can expect these areas—mainly North America, Western and Eastern Europe, Japan, Oceania, and the Soviet Union—to attain stable populations within about two generations. These stable populations will be approximately 30 to 60 percent larger than they are now, and this increase in numbers will add its quota of difficulties to the environmental problems facing mankind. But this ecological aspect of the human prospect, which we will examine later in this chapter, must be disentangled from the immediate problem of population overload.

The latter problem concerns the ability of those areas of the globe where population stability is not now in sight to sustain their impending populations even at the barest levels of subsistence. The dangers involved vary in intensity from nation to nation: there are a few areas of the underdeveloped world that are still underpopulated in their human carrying capacity. But in general the demographic situation of virtually all of Southeast Asia, large portions of Latin America, and parts of Africa portends a grim Malthusian outcome. Southeast Asia, for example, is growing at a rate that will double its numbers in less than 30 years; the African continent as a whole every 27 years; Latin

America every 24 years. Thus, whereas we can expect that the industrialized areas of the world will have to support roughly 1.4 to 1.7 billion people a century hence, the underdeveloped world, which today totals around 2.5 billion, will have to support something like 40 billion by that date if it continues to double its numbers approximately every quarter century.

Whether these horrific growth rates will in fact remain unchecked depends on two main variables. The first is the ability of the afflicted areas to introduce effective and stringent birth-control programs. Limited success in this regard has been enjoyed in a few places, mainly Taiwan and South Korea, although it should be noted that this "success" still leaves Korea and Taiwan among the fastest-growing populations in the world.[1] Almost no success has been attained in curbing growth rates in India or Egypt, despite official endorsement of population-control programs, and in those Latin American nations where growth rates are highest, population-control programs are not as yet even advocated. In fact, the only underdeveloped nation for which some cautious optimism may be voiced seems to be mainland China, where population-control programs, reportedly aimed at a zero growth rate by the year 2000, have been introduced with all the persuasive capability of a totalitarian educational and propaganda system.

Elsewhere in the underdeveloped world as a

1. See Kingsley Davis, "Population Policy: Will Current Programs Succeed?" *Science*, Nov. 10, 1967.

whole, population growth proceeds unhindered along its fatal course, with a virtual certainty of an 80 to 100 percent increase in numbers by the year 2000, and with projections thereafter that range between 6.5 billion and a grotesque 20 billion by the year 2050, depending mainly on estimates with regard to the rapidity of "spontaneous" or coerced changes in fertility.[2]

Still more alarming, even given the rise of governments of an efficiency and dedication comparable to that of China, a total curb on population growth appears to be impossible for the next century. This is because the fast-growing countries typically suffer from population age distributions in which almost half the population is below childbearing age. Therefore, even if drastic measures manage to limit families to a maximum of two children within a single generation, the steady advance of larger and larger numbers of individuals into their fertile years brings with it a vast potential increase in numbers. For example, if the underdeveloped countries were to achieve a zero population growth level of fertility by the year 2000, 50 years later they would nonetheless have increased in size two and a half times; if they succeed in achieving the target of "Western" fertility rates only by 2050, they will meanwhile have grown four and a half times in numbers.[3]

2. Tomas Frejka, "The Prospects for a Stationary World Population," *Scientific American*, March 1973.
3. *Ibid.*

For the next several generations therefore, even if effective population policies are introduced or a spontaneous decline in fertility due to urbanization takes effect, the main restraint on population growth in the underdeveloped areas is apt to be the Malthusian check of famine, disease, and the like. According to the 1967 report of the President's Science Advisory Panel on World Food Supply, malnutrition in the underdeveloped nations is already estimated to affect some 60 percent of their populations, with terrible costs in physical and mental retardation, while 20 percent suffer from undernourishment or actual slow starvation. All this contributes to preschool mortality rates three to forty times as high as those of the United States—a human tragedy of immense proportions, but also a demographic safety valve of great importance.

These Malthusian checks will exert even stronger braking effects as burgeoning populations in the poor nations press ever harder against food supplies that cannot keep abreast of incessant doublings. At the same time, the fact that their population "control" is likely to be achieved in the next generations mainly by premature deaths rather than by the massive adoption of contraception or a rapid spontaneous decline in fertility brings an added "danger" to the demographic outlook. This is the danger that the Malthusian check will be offset by a large increase in food production, which will enable additional hundreds of millions to reach childbearing age.

Here the situation hinges mainly on the prospects for the new "miracle" seeds, especially in rice and wheat, which have promised a doubling and tripling of yields. Fortunately or unfortunately, the future of the Green Revolution is still clouded in uncertainty. The new strains have not yet been adequately tested against susceptibility to disease, and there are suggestions from recent experience that they may be subject to blight. Perhaps more important in the long run is that all the new varieties of grains require heavy applications of water and of fertilizer. Water alone may be a serious constraint in many areas of the world; fertilizer is apt to prove a still more limiting one.

"Some perspective on this point is afforded," Paul Ehrlich writes, "by noting that, if India were to apply fertilizer as intensively as the Netherlands, Indian fertilizer needs alone would amount to nearly half the present world output."[4] Judging by the fact that of the 1.6 billion acres of currently cultivated land in the backward areas, less than 7 percent is now planted in the new seeds, a full "modernization" of agriculture would require enormous investments in fertilizer capacity. It is beyond dispute that these investments exceed by a vast margin the capabilities of the underdeveloped nations themselves, and it is possible that they exceed as well those of the devel-

4. Paul and Anne Ehrlich, *Population, Resources, Environment* (W. H. Freeman, 1972), p. 119.

oped world. More sobering yet, the introduction of fertilizers on such a scale may surpass the ecological tolerance of the soil to chemical additives.[5]

The race between food and mouths is perhaps the most dramatic and most highly publicized aspect of the population problem, but it is not necessarily the most immediately threatening. For the torrent of human growth imposes intolerable social strains on the economically backward regions, as well as hideous costs on their individual citizens. Among these social strains the most frightening is that of urban disorganization. Rapidly increasing populations in the rural areas of technologically static societies create unemployable surpluses of manpower that stream into the cities in search of work. In the underdeveloped world generally, cities are therefore growing at rates that cause them to double in ten years—in some cases in as little as six years. In many such cities unemployment has already reached levels of 25 percent, and it will inevitably rise as the city populace swells. The cesspool of Calcutta thus becomes more and more the image of urban degradation toward which the dynamics of population growth are pushing the poorest lands.

Only two outcomes are imaginable in this tragedy-laden historic drama. One is the descent of large portions of the underdeveloped world into a condition of

5. Barry Commoner, *The Closing Circle* (Knopf, 1971), pp. 84–93, *passim.*

steadily worsening social disorder, marked by shorter life expectancies, further stunting of physical and mental capabilities, political apathy intermingled with riots and pillaging when crops fail. Such societies would probably be ruled by dictatorial governments serving the interests of a small economic and military upper class and presiding over the rotting countryside with mixed resignation, indifference, and despair. This condition could continue for a considerable period, effectively removing these areas from the concern of the rest of the world and consigning the billions of their inhabitants to a human state comparable to that which we now glimpse in the worst regions of India or Pakistan.

But there is an alternative—and in the long run more probable—course of action that may avoid this dreadful "solution" to the overpopulation problem: the rise of governments capable of halting the descent into hell. It is certainly possible for a government with dedicated leadership, a well-organized and extensive party structure, and an absence of inhibitions with respect to the exercise of power to bring the population flood to a halt.

What is doubtful is that governments with such a degree of organization and penetration into the social structure will stop at birth control. A reorganization of agriculture, both technically and socially, the provision of employment by massive public works, and above all the resurrection of hope in a demoralized and apathetic people are logical next steps for any regime that is able to bring about social changes so fundamental as limitations in family size. The problem is, however, that these steps are likely to require a revolutionary government, not only because they will incur the opposition of those who benefit from the existing organization of society but also because only a revolutionary government is apt to have the determination to ram many needed changes, including birth control itself, down the throats of an uncomprehending and perhaps resistive peasantry.

Thus the eventual rise of "iron" governments, probably of a military-socialist cast, seems part of the prospect that must be faced when we seek to appraise the consequences of the population explosion in the underdeveloped world. Moreover, the emergence of such regimes carries implications of a far-reaching kind. Even the most corrupt governments of the underdeveloped world are aware of the ghastly resemblance of the world's present economic condition to an immense train, in which a few passengers, mainly in the advanced capitalist world, ride in first-class coaches, in conditions of comfort unimaginable to the enormously greater numbers crammed into the cattle cars that make up the bulk of the train's carriages.

To the governments of revolutionary regimes, however, the passengers in the first-class coaches not only ride at their ease but have decorated their compartments and enriched their lives by using the work and appropriating the resources of the masses who ride behind them. Such governments are not likely to view the vast difference between first class and cattle class with the forgiving eyes of their predecessors, and whereas their sense of historical injustice might be of little account in a world in which economic impotence also meant military impotence, it takes on entirely new dimensions in the coming decades for reasons connected with the changing technology of war. Thus a consideration of the population problem, as the first of the objective challenges of the human prospect, leads to an examination of the problem of war as the second of its imminent dangers.

What is new in the problem of war is, of course, the advent of nuclear weapons with their potential for "irreparable" damage, as contrasted with the much more restricted and more easily repaired damage of most conventional wars. As with the population problem, however, we are in danger of being rendered insensitive to the political ramifications of this element of danger in the human prospect by our tendency to picture it mainly in humanitarian terms.

The humanitarian aspect of nuclear war has focused our attention mainly on the stupendous killing power of the new weaponry. As Hans Bethe has described it:

Let us assume an H-bomb releasing 1,000 times as

much energy as the Hiroshima bomb. The radius of destruction by blast from a bomb increases as the cube root of the increase in the bomb's power. At Hiroshima the radius of severe destruction was one mile. So an H-bomb would cause almost complete destruction of buildings up to a radius of 10 miles. By the blast effect alone a single bomb could obliterate almost all of Greater New York or Moscow or London or any of the largest cities of the world. But this is not all; we must consider the heat effects. About 30 percent of the casualties in Hiroshima were caused by flash burns due to the intense burst of heat radiation from the bomb. Fatal burns were frequent up to distances of 4,000 to 5,000 feet. The radius of heat radiation increases with power at a higher rate than that of blast, namely by the square root of the power instead of the cube root. Thus the H-bomb would widen the range of fatal heat by a factor of 30; it would burn people to death over a radius of up to 20 miles or more. It is too easy to put down or read numbers without understanding them; one must visualize what it would mean if, for instance, Chicago with all its suburbs and most of their inhabitants were wiped out in a single flash.[6]

Our horrified fascination with these and similar statistics has led us to contemplate the consequences of nuclear warfare in terms of the obliterative results of using these weapons *en masse,* unleashing the 11,000 warheads now possessed by the United States or the 1,200 or so warheads possessed by the Soviets. Indeed, there are estimates of such an exchange, with

6. Hans A. Bethe, "The Hydrogen Bomb II," in *Scientific American Reader* (Simon & Schuster, 1953), pp. 194–95.

fatalities ranging from 50 to 135 million for the United States alone, depending on the defense "posture" of the various estimates.

It is understandable that we should be hypnotized by the vision of such ghastly possibilities. The risk, however, is that our concentration on this aspect of the consequences of nuclear warfare will lead us to overlook another result of the new technique of war. Essentially it resides in the fact that many small or relatively poor nations, even though they possess no fully developed industrial base or highly skilled labor force, can gain possession of nuclear weapons. As the example of China has shown, a nation with only a limited amount of industrial capacity can manufacture nuclear warheads by itself, although probably not missile delivery systems. The warheads can nonetheless be launched by bombers, smuggled into enemy harbors by ship, and so on. In addition, poor nations can obtain nuclear weapons as a by-product of the atomic power plants that many of them are now building or contemplating (or that will be built for them in the coming years by the developed countries).[7]

Thus there seems little doubt that some nuclear capability will be in the hands of the major underdeveloped nations, certainly within the next few decades and perhaps much sooner. The difficult question must then be faced as to how these nations might be tempted to use this weaponry. I will suggest that it

7. See Mason Willrich, "International Control of Civil Nuclear Power," *Bulletin of the Atomic Scientists*, May 1967.

may be used as an instrument of blackmail to force the developed world to undertake a massive transfer of wealth to the poverty-stricken world.

It may be, of course, that the governments of the underdeveloped world—and I would emphasize again the revolutionary cast of the governments that can be expected to arise in many places—will be able to arrange for the large-scale assistance they will need, and that they feel is owing to them, without recourse to these means. But given the reluctance to date of the developed world to offer more than token aid, and the likelihood that assistance on a scale large enough to raise the living standards of the six or eight billion of poverty-stricken inhabitants of the poor nations would necessitate an end to any advance, or even a decline, in the living standard of the well-to-do nations, the resort to ultimate tactics is surely not to be dismissed as a mere fantasy.

I do not raise the specter of international blackmail merely to indulge in the dubious sport of shocking the reader. It must be evident that competition for resources may also lead to aggression in the other, "normal" direction—that is, aggression by the rich nations against the poor. Yet two considerations give a new credibility to nuclear terrorism: *nuclear weaponry for the first time makes such action possible;* and *"wars of redistribution" may be the only way by which the poor nations can hope to remedy their condition.*

For if current projections of population growth rates are even roughly accurate, and if the environmental limitations on the growth of output, to which we will turn in our next section, begin to exert their negative influences within the next two generations, massive human deterioration in the backward areas can be avoided only by a redistribution of the world's output and energies on a scale immensely larger than anything that has hitherto been seriously contemplated. Under the best of circumstances such a redistribution would be exceedingly difficult to achieve. Given the constraints on economic growth that will make their presence felt with increasing severity, such an unprecedented international transfer seems impossible to imagine except under some kind of threat. The possibility must then be faced that the underdeveloped nations which have "nothing" to lose will point their nuclear pistols at the heads of the passengers in the first-class coaches who have everything to lose.

Beyond the outlook for a dangerous rise in international tensions for the reasons we have discussed, this "scenario" cannot be further developed with any degree of confidence. Richard Falk has rather melodramatically extended the plot as a series of increasingly grim "decades": the 1970s characterized by a Politics of Despair; the 1980s by a Politics of Desperation; the 1990s by a Politics of Catastrophe; the Twenty-first century as an Era of Annihilation.[8]

I take this macabre prophecy as the worst, not the

8. Richard Falk, *This Endangered Planet* (Random House, 1971), pp. 420f.

most likely, possibility. Even if nuclear blackmail is used, it need not lead to global disaster unless it resulted in an unleashing of nuclear conflict among the great powers. It is more plausible that a terrorist attack—for example, the wiping out of a city in an advanced nation that had refused to pay a ransom of a large portion of its material output—would serve as a stimulus to bring a substantial reduction in nuclear armaments coupled with worldwide nuclear inspections, especially in the "dangerous" underdeveloped countries. Such a protective reaction would not reduce the chances for conventional limited wars—indeed, it might even increase them—but it would greatly reduce the risk of further nuclear threats of the kind we have described.

Unfortunately, even this happiest outcome to the immediate risks of nuclear catastrophe will not remove the influence of war as a fundamental molding element in the human prospect. As we have said, the danger of "limited" war remains, and the probability of such wars is very high. The frequency of "deadly quarrels" showed no signs of decline over the two centuries before 1940,[9] and experience in the past three decades is hardly encouraging: a casually assembled list includes civil conflicts in Greece, Korea, Nigeria, Pakistan, Indonesia, Sudan, and, on a smaller scale, Ireland; minor international sorties led by India,

9. L. F. Richardson, "The Statistics of Deadly Quarrels," in *The World of Mathematics* (Simon & Schuster, 1956), II, 1254.

Pakistan, England, France, Egypt, Israel, Portugal, China, and North Korea; major invasions conducted by the Soviet Union and the United States. Very probably wars on this scale, with this frequency of occurrence, will continue as long as nation-states continue to play their role as the main forms of mass social organization.

It is this last point that is of the essence. The continuing likelihood of war enters the human prospect not alone by virtue of the life-or-death risks it offers, but also as a principal reason for the continuation of nation-states as the dominant mode of social organization. The latter, in turn, gives unhappy assurance that nationalism, with all its potential for historic calamity, will be encouraged by the persisting realities of international existence—the omnipresent threat of war justifying the need for nation-states; the presence of nation-states in turn setting the stage for a continuance of the threat of war. From this vicious circle there is at present no escape, a fact that sets severe limits, as we shall see, on what one can expect by way of a fundamental response to many of the challenges ahead.

We shall return to the problem of the nation-state more than once in our subsequent chapters. But we cannot conclude this examination of the external dangers facing mankind without adding a third problem to those of population growth and war. This is the danger, to which we have already alluded, of encroaching on the environment beyond its ability to support the demands made on it.

Here we come to a crucial stage of our inquiry. For unlike the threats posed by population growth or war, there is an ultimate certitude about the problem of environmental deterioration that places it in a different category from the dangers we have previously examined. Nuclear attacks may be indefinitely avoided; population growth may be stabilized; but ultimately there is an absolute limit to the ability of the earth to support or tolerate the process of industrial activity, and there is reason to believe that we are now moving toward that limit very rapidly.

When we examine the actual timetable of environmental disruption, however, we soon encounter a baffling set of considerations. Despite the certainty of our knowledge that a limit to growth impends, we have only a very imprecise capability of predicting the time span within which we will have to adjust to that impassable barrier. As we shall see, this makes it difficult to formulate appropriate policies, or to forecast the rate of social change that will be required to bring about the necessary environmental safeguards.

Take, as our initial problem, the availability of the resources necessary to sustain industrial output. In the developed world, industrial production has been growing at a rate of about 7 percent a year, thereby doubling every ten years. If we project this growth

47

rate for another fifty years, it would follow that the demand for resources would have doubled five times, requiring a volume of resource extraction thirty-two times larger than today's; and if we look ahead over the ten doublings of a century, the amount of annual resource requirements would have increased by over a thousand times.

Do we have the resources to permit us to attain—or sustain—such gargantuan increases in output? Here the problem begins to reveal its complexity. A considerable proportion of the resources we extract today does not become industrial output but ends up as waste. To the extent that we can reduce waste, or use old outputs as new inputs—for example, recycling junked cars as new steel—we will be able to reduce the need for new resources, although by how much no one knows. Further, the problem is complicated because we are largely ignorant of the extent of most of the world's resources, petroleum being perhaps an exception. Indeed, not only is the world still largely "unexplored," so far as its potential mineral and other riches are concerned, but the very definition of a resource changes as our ability to extract minerals or other substances improves. For example, today we utilize enormous reservoirs of iron ore that were not even considered to be reserves when we were still mining the rich iron deposits of the Mesabi Range, now long exhausted. In point of fact, reserves of all known elements exist in "limitless" quantities as trace elements in granite or sea water, so that, given the appropriate technology and the availability of sufficient energy, no insurmountable barrier to growth need arise from resource exhaustion for millennia to come.

This conclusion depends, however, on several assumptions. It assumes that we will develop the necessary technology to refine granite or sea water before we run out of, say, "copper"—meaning copper in its present degree of availability.[10] More important yet, it assumes that the ecological side effects of extracting and processing the necessary vast quantities of rock or sea water would not be so deleterious as to rule out the new extraction technologies because of their environmental impact. Most important of all, as we shall see, the gigantic energy requirements for mining ordinary rocks or refining sea water bring us to the consideration of whether a continuously increasing application of energy is compatible with environmental safety.

To many of these questions no clear-cut answers exist. We do not know how rapidly new technologies of extraction or refining can be developed, or the degree to which anti-pollution technologies can suppress their ecological disturbance. Today, for example, the practical limit to open-pit mining, which appears to be the most economical way to extract

10. See T. S. Lovering, "Non-Fuel Mineral Resources in the Next Century," in *Global Ecology*, eds. John P. Holdren and Paul Ehrlich (Harcourt, Brace, Jovanovich, 1971); and Preston Cloud, "Mineral Resources in Fact and Fancy," in *Toward a Steady-State Economy*, ed. Herman Daly (W. H. Freeman, 1971).

common rock, is about 1,500 feet. It seems unlikely that this depth can be doubled, and it is a certainty that the rock extracted from such a vast pit will diminish exponentially unless ways can be found to dig pits with vertical walls.[11] In addition, as T. S. Lovering has written, "The enormous quantities of unusable waste produced for each ton of metal are more easily disposed of on a blueprint than in the field."[12]

But even if we make the heroic assumption that all these difficulties will be overcome, so that another century of uninterrupted industrial growth, with its thousandfold increase in required inputs, will face no constraints from resource shortages, there remains one barrier that confronts us with all the force of an ultimatum from nature. It is that all industrial production, including, of course, the extraction of resources, requires the use of energy, and that all energy, including that generated from natural processes such as wind power or solar radiation, is inextricably involved with the emission of heat.

The limit on industrial growth therefore depends in the end on the tolerance of the ecosphere for the absorption of heat. Here we must distinguish between the amount of heat that enters the atmosphere from the sun or from the earth, and the amount of heat we *add* to that natural and unalterable flow of energy by man-made heat-producing activities, such as industrial

11. Cloud, *op. cit.*, p. 61.
12. Lovering, *op. cit.*, p. 45.

combustion or nuclear power. Today the amount of heat added to the natural flow of solar and planetary heat is estimated at about 1/15,000 of the latter—an insignificant amount.[13] The emission of man-made heat is, however, growing exponentially, as both cause and consequence of industrial growth. This leads us to face the incompatibility of a fixed "receptacle," however large, and an exponentially growing body, however initially small. According to the calculations of Robert Ayres and Allen Kneese, of Resources for the Future, we therefore confront the following danger:

Present emission of energy is about 1/15,000 of the absorbed solar flux. But if the present rate of growth continued for 250 years emissions would reach 100% of the absorbed solar flux. The resulting increase in the earth's temperature would be about 50° C.—a condition totally unsuitable for human habitation.[14]

Two hundred and fifty years seems to give us ample time to find "solutions" to this danger. But the seemingly extended timetable conceals the gravity of the problem. Let us suppose that the rate of increase in energy use is about 4 percent per annum, the worldwide average since World War II. At a 4 percent rate

13. W. R. Frisken, "Extended Industrial Revolution and Climate Change," E⊕S, American Geophysical Union, vol. 52 (July 1971), p. 505.

14. Robert U. Ayres and Allen V. Kneese, *Economic and Ecological Effects of a Stationary State*, Resources for the Future, Reprint No. 99, December 1972, p. 16. See also Frisken, *op. cit.*, and John P. Holdren, "Global Thermal Pollution," in *Global Ecology*.

of growth, energy use will double roughly every eighteen years. This would allow us to proceed along our present course for about 150 years before the atmosphere would begin to warm up appreciably—let us say by about three degrees. At this point, however, the enormous multiplicative effects of further exponential growth would suddenly descend upon us. For beyond that threshold, extinction beckons if exponential growth continues for only another generation or two. Growth would therefore have to come to an immediate halt. Indeed, once we approached the threshold of a "noticeable" change in climate, even the *maintenance* of a given industrial level of activity might pour dangerous amounts of man-made heat into the atmosphere, necessitating a deliberate cutting back in energy use.

In point of fact, serious climatic problems may be encountered well before that dangerous threshold. Noticeable perturbations are anticipated by climatologists when global man-made heat emissions reach only 1 percent of the solar flux, little more than a century from now.[15] This timetable assumes, however, that the rate of energy dissipation will not rise from its present rate of annual increase of 4 percent to, say, 5 percent or even higher. These estimates therefore make no allowance for *increases* in the rate of global heat dissipation if massive industrialization is under-

15. Frisken, *op. cit.*, p. 505.

taken in the underdeveloped regions. Per capita energy consumption in these areas is now only about one-tenth of that of the more advanced portions of the globe, although populations in the backward regions outnumber populations of the industrialized areas by two or three times. To raise per capita energy consumption in the poor regions of the world to Western levels would therefore require a twenty- to thirty-fold increase in energy use in these areas—a calculation that, however staggering, still fails to take into account the potential demands for energy from populations, within these areas, that will certainly double and possibly quadruple over the next hundred years.

It is important, in considering this last element of the human prospect, to avoid a prediction of imminent disaster. The timetable for global climatic disturbance is not only fairly distant, as we are accustomed to judge the time scale of events, but it can be pushed still farther into the future. Increases in the efficiency of power generation or utilization may considerably augment the amount of industrial production obtainable per unit of energy. New technologies, above all the use of solar energy, which adds nothing to the heat of the atmosphere since it utilizes energy that would in any case impinge on the earth, may greatly reduce the need to rely on man-made energy. From yet a different perspective, the technologies required to supplant the present fossil fuels—safe and efficient fission reactors, economical solar or wind machines, large-scale geo-thermal plants—may not arrive "on time," thereby enforcing a slowdown in the rate of energy use and postponing the advent of an ecological Armageddon. More important, the vast energy sources required to "melt the rocks and mine the seas," notably fusion power, may also remain beyond our capability for a very long period, thereby curbing our fatal growth curve by depriving us of the needed resources. Finally, a wholesale shift away from material production to the production of "services" that demand far less energy would also greatly extend the period of safety—a possibility that we will look into in our next chapter.

Thus imminent disaster is not the problem here. It is the inescapable need to limit industrial growth that emerges as the central challenge. Indeed, the main lesson of the heat problem is simply to drive home with the greatest possible force the conclusion that such a limitation must sooner or later impose a strait-jacket on the never-ending growth of industrial production, even under the most optimistic or unrealistic assumptions with regard to resource availability or technology.[16]

The problem of global thermal pollution, for all its

16. I must add a footnote here, lest it be thought that the availability of safe solar energy obviates the problem of an energy constraint. Ayres and Kneese (page 51) point out that 250 years of growth, with its present associated emission of heat, would reach 100 percent of the total solar flux.

awesome finality, therefore stands as a warning rather than as an immediate challenge. Difficulties of a much more matter-of-fact kind—resource availability, energy shortages, the pollution resulting from noxious by-products of industrial production—are likely to exert their throttling effect long before a fatal, impassable barrier of irreversible climatic damage is reached. Every sign, however, points in the same direction: industrial growth must surely slacken and likely come to a halt, in all probability long before the climatic danger zone is reached.

Once again, however, we must stress an aspect of the environmental problem that is largely overlooked in the mounting literature on the ecological threat. Most of this literature focuses on the technical aspects of the problem, whose dimensions we have genera'ly described. Of far greater importance for the human prospect are its socio-economic and political consequences. It is these aspects which will therefore mainly occupy us in the chapters to come.

There remains one concluding comment, before we proceed. At the outset I said that three elements of the current human predicament would be unanimously

---

It follows, therefore, that even the fantasy of a complete capture of all sunlight falling on the earth would yield no more energy than 250 years of growth of conventional (including nuclear) sources. Beyond that lies the exotic possibility of capturing additional solar energy in space and safely relaying it to earth by microwaves, or using microwaves to radiate man-made energy into space. The substantial application of such technologies seems far beyond any realistic capabilities of the next century or so.

selected if we were to seek the source of the pervasive unease of our contemporary mood. Now, without going beyond the specific dangers of population growth, war, and environmental deterioration, I must identify a fundamental element in the external situation—not so much a fourth independent threat as an unmentioned challenge that lies behind and within all of the particular dangers we have singled out for examination. This is the presence of science and technology as the driving forces of our age.

It is hardly necessary, I think, to spend much time defending the cogency of this unifying proposition. The population explosion that looms with such horrifying possibilities is directly traceable to the consequences of new techniques of science and technology in the area of medicine and public health: it is not a rise in fertility rates but a science-induced fall in death rates that has set off the unstable demographic situation that now threatens to overwhelm the underdeveloped areas. The responsibility of science and technology for nuclear armaments is self-evident, as is also their joint effect in bringing about both the rate of industrial expansion and the peculiarly dangerous nature of modern industrial processes. That science and technology may also be indispensable agents for the mitigation of these external dangers, through birth-control techniques, sophisticated means of arms detection or defense, or greatly improved methods of energy production and pollution suppression, does not vitiate the contention that these external dangers arise in the first instance because of the development of science and technology in that era we call "modern history."

The very possibility of using science and technology to mitigate our present problems indicates, however, that it is not the extraordinary development of these forces, as such, that underlies our predicament. It is, rather, their fusion in a civilization that has developed scientific technology in a lopsided manner, giving vent to its disequilibrating or perilous aspects without matching these ill effects with compensating "benign" technologies or adequate control mechanisms. In turn, this raises the question of whether scientific research and technological application follow their "own" courses of development, or whether these forces are imperfectly constrained and directed because of inadequacies of the economic and social milieu within which they have arisen.

That is a question for our next chapter. Here it is enough to claim that the external challenge of the human prospect, with its threats of runaway populations, obliterative war, and potential environmental collapse, can be seen as an extended and growing crisis induced by the advent of a command over natural processes and forces that far exceeds the reach of our present mechanisms of social control. It goes without saying that this unequal balance between power and control enters into, or provides the underlying

57

basis for, that "civilizational malaise" of which I spoke earlier, and to which we will return.

At this point, however, we have still to push beyond the facts, as we can best identify and interpret them, to an analysis of their full impact on the human prospect. We have identified the external challenges; what remains to be examined is the response that can be mustered against these challenges—a response that now appears not alone in curbing or avoiding the specific threats we have mentioned but in coping with the dangerous tendencies of industrial civilization itself.

AFTERWORD

ARE THE EXTERNAL CHALLENGES still so grim? I think so. Of course, the "demolition of research" has undermined some of my data, and events have forced me to alter my thinking in part about the implications of those data that remain. Yet, the thrust of the argument, were it to be written today, would depart in detail but not in direction from the preceding pages. The external setting within which the human prospect will unfold seems to me as difficult and threatening as when I first took its measure, although (as I have commented earlier), I think we are less responsive today to the message it conveys.

Nonetheless, changes have taken place and a good way to begin this reconsideration is to call attention to them. The first has to do with the altered outlook for long-term population growth. It is curious that this great force in human affairs, which so much resembles an impersonal process, has again and again proved recalcitrant to accurate prognostication. When I began to write the *Inquiry* in 1972, prevailing expert opinion was virtually unanimous in endorsing the views I set forth. Within the last few years it has altered greatly.

For one thing, it is now widely agreed that a rapid fall in birthrates will bring population stability to the advanced countries much more quickly than the "two generations" that I anticipated in the text. In many Western countries, the net reproduction rate has already gone below replacement levels—that is, mothers are not replenishing their own numbers. Of course it will take twenty-some years before the current crop of female babies reaches childbearing age, and fashions or laws or technology may greatly alter their desire for offspring. But there is no reason to *expect* that to be the case, although population movements may fool us again, as they have in the past.

Zero Population Growth (ZPG) has been approaching for a long time in the West. Part of the reason is social and technological—the widening adoption of simple birth control methods. Part of the reason is economic—the increasing expense of children in an urban environment with small nuclear families. Part of the reason may be psychological or even moral—a kind of indictment of a civilization in which children become "nuisances" and "burdens." The fall in the birthrate is therefore the outcome of many forces, perhaps including that civilizational malaise of which I have spoken. Why that fall has taken place during the last decade more rapidly than expected we do not know.

Whatever the ultimate causes, there is no doubting the phenomenon. In West Germany, East Germany, Austria, and Luxembourg, birthrates have declined so far that these countries are already experiencing a slight actual shrinkage in their numbers. In the United States and the rest of Europe and Japan, the situation has not yet reached Negative Population Growth (NPG), but it is approaching ZPG. In the United States, for example, we are now thought to be at ZPG so far as native Americans are concerned, although our population totals are still swollen by some 400,000 annual legal and illegal immigrants. Demographers predict that our population will grow from today's 220 million to about 250 million by the year 2020, thereafter leveling off.

To many people, ZPG or NPG still carry negative overtones. It is hard to shake the association between a growing population and a healthy society (and insofar as NPG is a vote of no confidence in our way of life, the judgment is right). But the end of the population "problem" in the West has important positive consequences for the human prospect. It lessens our need to exploit the world's resources; it reduces our pressure against the environment; it throttles back on the required rate of growth. All things considered, it is a favorable development, and one that was not to be expected so quickly when I first took the measure of the population issue.

But the really remarkable change has occurred in the underdeveloped world, where the population outlook was most dire. *During the five-year period 1970–1974, the trend in world population turned down for the first time in recorded history.* The figures do not sound dramatic: in 1970 the rate of world population increase was 1.90 per year; in 1975 it was 1.64. The actual growth in numbers of people in 1970 was 69 millions; in 1975, 64 millions. Note that nothing like ZPG is at hand on a planetary scale. World population is still rising. But at least—and at last—the rate of increase has tapered off.

Further, there is reason to expect that it will continue to taper off. For the decline in global growth rates is due only in small degree to the drop in birthrates in the Western world. It is mainly the consequence of a long-awaited, long-despaired of, fall in birthrates in the nonindustrialized world. Here the single biggest factor has been the success of Chinese birth control policy. If we can believe their population statistics, the Chinese birthrate has fallen from 32 to 19 per 1,000, a gigantic accomplishment and one that makes a tremendous difference to global projections because China represents a fifth of the world's population.[1]

Moreover, world population growth will probably continue to taper off, even if the Chinese miracle turns out to be less miraculous than reported. Birthrates are also falling in many Latin American countries, such as Mexico and Brazil, because the long-standing opposition of church and state to family planning has finally turned into an endorsement of birth control, now admitted to be an essential precondition for ending squalor and misery. In India, efforts to curb births have gone so far as to include the advocacy of compulsory sterilization in certain cases—a policy that backfired badly for Indira Gandhi, who proposed it.

As a result of these and similar efforts, global demographers now project that world numbers will rise from their present estimated 4 billions (400 millions more than when I wrote my chapter) to a possible leveling-off of eight to ten billions by the middle of the next century.[2] These are still enormous numbers, but we must recognize the victory they represent. The trend is now toward a manageable plateau, not toward

1. For a skeptical assessment of Chinese demography, see Nick Eberstadt, "Has China Failed?" *New York Review of Books*, April 5, 1979. Eberstadt believes that the fall in population growth rates is probably less than the figures I have used, which come from *Population Trends; Signs of Hope, Signs of Stress*, Worldwatch Paper no. 8 (Washington, D.C.), October 1976.

2. Donald Bogue and Amy Ong Tsu, "Zero World Population Growth?" *The Public Interest*, Spring 1979. For a recent review of population prospects, see also *Resource Trends and Population Policy*, Worldwatch Paper no. 29 (Washington, D.C.), May 1979.

a hopeless and endless proliferation. A long-term outlook of cautious hopefulness has displaced the nearly universal gloom that prevailed when I surveyed the field in the early 1970s.

Unfortunately, that is not quite the end of the problem. If the cancer is now spreading less rapidly, it is still spreading and will continue to spread for many decades. The ravages of overpopulation will still be experienced in many parts of the world during our lifetimes and the lifetimes of our children, bringing us the horrors of famine and malnutrition and poverty. We have had several serious famines since *The Human Prospect* first appeared; one in Bangladesh, one in Ethiopia, one in Sahel, the borderland region just south of the southward creeping Sahara Desert. In that Sahelian disaster alone, between 100,000 and 250,000 people perished of hunger. Many Americans saw on their TV screens the skeletonlike goats and cattle of the region being driven southward by equally skeletonlike herders, in search of grazing land and food.

Thus, despite the long-term declining rate of growth, the short-term pressures of population still push many areas of the world beyond the brink of their sustaining powers. Consider land. About half of the world's potentially arable land is under cultivation today. It is, of course, the best half. To put the remaining half under the plough will require immense inputs of irrigation, fertilizer, and other equipment. The

Food and Agriculture Organization has projected that we would need to treble our investments in irrigation alone to bring about a mere 20 percent increase in arable land area. Quite aside from the overwhelming demands that this would place against the capital building capacities of the underdeveloped world, it would also require a 250 percent increase in the use of water, itself a scarce resource in many food-poor areas. Fertilizers, essential for any significant increase in output, have been in worldwide short supply since 1971: the price of phosphate rock from Morocco, a main source of fertilizer, has quadrupled. And energy, the key to increases in agricultural as well as industrial output, remains a critical bottleneck—more critical, as we shall see, than it was eight years ago.[3]

In sum, although we are within sight of ZPG as a long-term global possibility, we are not within sight of zero hunger or zero desperation during the time span in which our own human prospects will be lived out. The pressure of mouths against food resources will continue for at least the rest of the century, and possibly beyond that. Hence the stage is still set for the political consequences of hunger—consequences that led me, as the reader knows, to raise the possibility of iron governments and wars of redistribution as lurking outcomes for the future. Let us therefore turn to these problems of war and strife.

3. For the above data, see *Faith, Science and the Future*, World Council of Churches, Geneva, 1978, pp. 125f.

I am not certain whether the events of the last years have lent plausibility to my disquieting scenarios or not. We have seen the Sahelian disaster come and go, for example, leaving nothing but corpses, not a revolutionary government. It may be that the outcome of the overcrowding and forced rural exodus we find in Asia or Latin America will simply be a long-lasting condition of demoralization and social disintegration, without the social rekindling that I used to think a probable outcome. If that is the case, millions will perish or languish, but the well-being of the richer portion of the world will be the beneficiary of their sufferings.

I would not want to count on a lessening of international tensions, however. Unhappily, there is far more evidence today than when I wrote the text that nuclear weaponry is available by clandestine means or by direct assembly to any nation (perhaps even to terrorist groups) with a few tens of millions of dollars. The purely technical feasibility of nuclear war or blackmail is therefore appreciably greater than when I examined the subject in 1972, and by all accounts will be greater still ten years hence. I do not know what conclusion to draw from this other than that the increased risks must be recognized and lived with as best we can.

Meanwhile, the means of international conflict have changed dramatically in another respect. For the poorer nations—rather, a lucky few among them—

now possess a weapon that they did not have in 1972, namely their united control over the flow of oil. The rise of the Organization of Petroleum Exporting Countries (OPEC) has been a tremendous event in altering the balance of world economic power, giving rise to the largest bonanza in history—greater than the gold that enriched Spain in the sixteenth century or the wealth of the Indies that England amassed in the eighteenth and nineteenth centuries. In 1979 alone, OPEC net revenues—the surplus of receipts over expenditures abroad—will add $130 billions to the coffers of the members of the cartel.

The consequences of this totally unforeseen diversion have been immense. The advent of "oil shock" precipitated a substantial recession throughout the Western world, revealing the dependence of the West on the oil producers. The reduced availability of cheap energy has cut the rate of economic growth in the industrial world, hastening the "limits to growth" scenario to which my analysis pointed. And vast riches have accrued to the oil sheikdoms, which have created a kind of welfare feudalism at home and an impressive portfolio of capitalist investments abroad.

Thus OPEC has driven a wedge between the oil-producing nations and the industrialized world, giving rise to stories about the conditions under which the United States might seek to secure Arabian oil by military intervention, or at the other extreme, to the "take over" of Western capitalism by Arabian sheiks. In fact American military and Arabian financial occupation are both fantasies. What we have, rather, is a new arena in which the tensions arising from the uneven development of the world are made manifest—an arena filled with potential dangers for both sides.

It is interesting to speculate whether cartels similar to OPEC could be created for other raw materials, such as bauxite, taconite, manganese, and the like. As far as we can see, the political and geographical distribution of these resources does not lend itself to cartelization as in the case of oil. Moreover, opposed to the actual power of OPEC and to the possibility of other cartels is the steady accumulation of "food power" into the hands of the United States and Canada. Each year the Asian and African nations go more deeply into "food deficit," as they import more and more grains from the North American basket areas to feed their populations. Thus economic power does not lie wholly on one side, and the outcome of the tension between the rich and the poor nations has become less easy to predict, with each side the hostage of the other.

Overall, I would say that the hopes for a marked improvement in the world's lopsided material condition are still very small, despite the bonanza of oil. Even here, the experience of the oil-rich nations has shown how little those riches can mean for general industrialization and modernization, despite the wealth bestowed on a ruling elite, and the glitter of a capital with Hilton hotels and modern office buildings. The world is still hopelessly split into areas of wealth and poverty, with little prospect of a narrowing of that gap. The politics of international economic affairs in our lifetimes must therefore be a politics of inequality, inherently a politics of mutual suspicion and struggle.

The crucial element today, as eight years ago, remains the environment—the ability of the planet to sustain the mushrooming of industrial output and to absorb the destruction that is the consequence of that vast human effort.

Eight years have not brought significant change to my assessment of these possibilities and dangers. To be sure, a vast field of oil has been discovered in Mexico that may prove ultimately to rival that of Saudi Arabia. Enormous beds of manganese nodules have been found on the seafloor. Estimates of the "crustal abundance" of the earth's primary minerals give us staggeringly large volumes of latent resources that can be used as mutually interlocking substitutes. Thus the prospects of physical resource availability are somewhat more reassuring than they were eight years ago, in some degree because the awareness of an impending pinch has prompted worldwide exploration and technological improvement.

That is not a full reassessment of the picture, however. The key to the resource problem, today as before, is energy; and energy continues to mean oil, at least for the next ten to twenty years. And despite the great Mexican oil discovery, we have a virtual consensus that oil is a fast disappearing resource. Under almost any set of realistic supply and demand assumptions, we have only enough oil to sustain our present (reduced) rate of growth for roughly 25–50 years.[4] And the technical and safety outlook for a rapid switch to nuclear power is worse, not better, than we expected only ten years ago.

As before, it is the exponential character of growth that poses the challenge. Take coal, the most abundant energy resource for the United States. If coal consumption in the United States were to continue at the levels of the 1970s, our coal reserves (1.49×10^{12} metric tons) would last several thousand years. But if our rate of use of coal were to rise by only one percent each year, our stocks would be used up in 342 years. With a 7 percent annual increase in use, they would be gone in 80 years![5]

The power of exponential growth comes home even more forcefully with regard to oil. It is now believed that untapped petroleum reserves amount to about one and a half to two trillion barrels. This is enough to support 100 years of consumption at today's rate of use. But if oil consumption grows only at the modest rate of 3 percent per year, the reservoir of oil

4. See Andrew Flower, "World Oil Production," *Scientific American*, March 1978, and Charles Issawi, "The 1973 Oil Crisis and After," *Journal of Post-Keynesian Economics*, vol. 1, no. 2 (Winter 1978–79).
5. *Faith, Science and the Future, op. cit.*, p. 135.

shrinks to a mere fifty years' consumption. Moreover, even a doubling of the estimates of oil reserves—far in excess of the most optimistic hopes—adds only *twenty* years of supply, so long as consumption mounts by 3 percent each year.[6]

None of these and similar statistics imply that we will run out of resources or mine away the planet until nothing remains beneath our feet. Their implications are the same as eight years ago—namely, *that the rate of industrial growth will have to accommodate itself to the increasing difficulty of mining or refining reserves. It is not a sudden stop that impends; it is a gradual slowdown.*

As we have mentioned, that slowdown has already appeared. Growth rates throughout the world are down by a third to a quarter, if we compare the years 1975–1979 to the period 1970–1974. To be sure, this slowdown is labeled "recession" and is blamed on such factors as the OPEC rise in oil prices. But *"recession" is precisely the form that a slowdown in growth takes.* The problem then lies in the political and social repercussions that recession might bring, a matter to be considered in the next chapter.

Does our current, already rather protracted, recession give us a foretaste of what is to come? To a limited extent, this is undoubtedly the case. The soaring price of oil and oil products has certainly pulled the

6. Charles Komanoff, "Doing Without Nuclear Power," *New York Review of Books*, May 17, 1979, p. 14.

economy backward and reduced our real incomes. But the squeeze is still very mild. Our pre-OPEC use of oil was profligate, so that very large possibilities for conservation are available to us. We are already producing 50 percent more GNP per barrel of oil, and can produce substantially more, if (and as rapidly as) we redesign our automobiles to get 30 and 40, not 13 and 14, miles per gallon, or as we build buildings to use solar, not oil heat, or as we simply learn to turn off lights and turn down thermostats.

Thus energy constraints do not yet imply any serious reduction in absolute living standards, only a new set of economic pressures. For the short run, we are entering a period of tightening energy supplies that will have mixed effects—partly impairing our ability to run the economy effectively, partly serving as a stimulus to alter our technological structure, and partly as an excuse to ease up on anti-pollution measures. These are all serious problems, but they are not unmanageable ones.

It is when we turn to the long-term problems that we encounter the truly daunting elements of the human prospect. Here I see little need to alter the thrust of my original argument, even though, since the first edition appeared, the attention of scientists has been directed at the climate problem from a somewhat different perspective than its long-term heating-up from the release of combustion energy. The emphasis today is on a short-term effect that results from the release of carbon dioxide into the atmosphere as a byproduct of combustion. There, the CO_2 forms an invisible "pane" of gas that acts like the glass in a greenhouse, trapping the reflected rays of the sun, and warming the atmosphere just like the air in a greenhouse.

According to current scientific estimates, the amount of CO_2 in the air is expected to double by the year 2020. This addition to the "window pane" within our atmosphere would be sufficient to raise surface temperatures on earth by some 1.5° to 3.0°. From this seemingly small change in temperatures, far-reaching results would ensue. The difference between average global temperatures during the last Ice Age and those prevailing today is only about 5°, but that difference was enough to have raised the sea level by 100 meters since the Ice Age because enormous quantities of water, formerly locked up as ice, were released into the oceans. Among the projected effects of a rise in temperatures of 1° to 3° in our planetary greenhouse would be the further unlocking of vast amounts of water still congealed in our polar ice caps. This could eventually bring sea levels above the level of the land in the populous delta areas of Asia, the coastal areas of Europe, and much of Florida.[7] Long before that it is feared that the rise in temperature would have irreversibly altered rainfall patterns, with grave potential effects.

Much of the greenhouse effect remains unclear.

7. *Pollution, the Neglected Dimension*, Worldwatch Paper no. 27 (Washington, D.C.), March 1979.

But the scientific community is sufficiently alarmed to have called a worldwide conference on climate problems, and to issue warnings couched in near-peremptory language. In its 1977 study, *Energy and Climate,* the U.S. National Academy of Science concluded that climatic dangers might require the phasing out of fossil fuels within the next fifty years.

Thus the looming, ill-defined problem of overloading the environment remains as real, but as indistinct, as when *The Prospect* was first written. Moreover, one thing has become clearer since then, admitted by those who are congenital optimists as well as by those who are not. It is that the sheer scale of our intervention into the fragile biosphere is now so great that we are forced to proceed with great caution lest we inadvertently bring about environmental damage of an intolerable sort. An instance in case was the debate, shortly after the publication of *The Prospect,* over the potentially harmful effects of the fluorocarbons released into the atmosphere as the propellant gas in hair sprays, underarm deodorants, and the like. What was striking about this debate was not its outcome, which barred the use of fluorocarbons, but the admission by foes as well as friends of the investigation, that the spewed-out tonnage of these trivial products was large enough to alter the filtration properties of the atmosphere.

This staggering and wholly unprecedented scale of invasion into the biosphere is the element of the external challenges that becomes, in my estimation, of central importance for the medium-term future. For the threat is not merely that of disposing of a few tons of nuclear waste, but of hundreds and eventually thousands of tons; not just that of controlling chemical pollution in a lake or a river, but in the seas and the atmosphere as a whole; not just coping with a temperature inversion over Los Angeles or New York, but with climatic changes that would affect continental rainfalls, monsoons, and growing seasons. This massive assault against the biosphere becomes, along with the availability of energy, the main determinant of our potential rate of industrial expansion, and thereby of our ability to continue our present mode of production more or less unchanged.

In the final Afterword of this book I will make some rough calculations as to when this period of enforced change might begin. But at this point, while I am still summing up and reconsidering the nature of the external challenges themselves, it is sufficient to establish, with as much certainty as contemporary knowledge permits, that the threats are still there, imperious and unavoidable.

Thus the changes in known facts do not significantly affect the general conclusions at which I arrived almost a decade ago. And on reconsideration that is hardly surprising. For the ultimate "external" challenge is nothing but the forces of science and technology—those djinns of modern man who have been released from their bottles and now defy any attempts to put them back. The true measure of the external threat is the human capacity to alter, irrevocably and perhaps disastrously, the conditions of its own existence. That capacity, which humankind has always possessed in a modest degree, has been concentrated and magnified in our age, until it opposes mankind as an alien force, seemingly not of its own making. The pressure of population with its capacity for social disorder; the terrific armamentarium of weaponry and now the new strategic weapon of energy itself; the ceaselessly turning, ever faster-spinning industrial mechanism with its ever-growing inputs of minerals and fuels and its ever-growing outputs of material and caloric wastes—these are the final challenges of our time, challenges that cannot possibly be removed without the most profound changes in our industrial and scientific underpinnings.

That is why the knowledge that I have drawn on to describe the condition of mankind may change in detail—even in very far-reaching detail—without changing the conclusion to which such an inquiry brings us. The external challenges may escape our sight these days because we are preoccupied with other things, but they are there, the products of our own civilization, and they must be reckoned with if we are to think seriously about the prospects for that civilization.

THREE

Socio-Economic Capabilities for Response

OUR LAST CHAPTER laid out for examination the external dangers of the human prospect. Yet that list of dangers still does not fully describe the challenge of the human prospect, nor wholly account for the somber state of mind with which we look to the future. For the dangers we have discussed do not descend, as it were, from the heavens, menacing humanity with the implacable fate that would be the consequence of the sudden arrival of a new Ice Age or the announcement of the impending extinction of the sun.

On the contrary, as we have repeatedly sought to emphasize, all the dangers we have examined—population growth, war, environmental damage, scientific technology—are *social* problems, originating in human behavior and capable of amelioration by the alteration of that behavior. Thus the full measure of the human prospect must go beyond an appraisal of the seriousness of these problems to an estimate of the likelihood of mounting a response adequate to them, and not least to some consideration of the price that may have to be paid to muster such a response.

The question is where to begin. Immediately two possibilities appear. The first is to discuss the question of adaptation and response in terms of our *individual* capabilities for change. This is an approach at once superficial and profound—superficial if it only leads us in the direction of moral exhortation or admonitions to change our behavior according to the dictates of reason; profound if it brings us to reflect on the ultimate capacities for individual change that may be rooted in what we call "human nature." The second possibility is to discuss the problem of response in terms of the flexibility of the social organizations that mobilize human effort and that powerfully influence human activity, in particular those massive social instruments for shaping behavior we call nation-states and economic systems.

I propose that we take the second avenue of analysis before the first, concerning ourselves in this chapter with generalizations and speculations about our collective capacity for response, and reserving to the next chapter various reflections on the problem of human nature.

This still leaves us with a choice of procedures. Shall we begin our consideration of the social capacity for action with an analysis of nation-states or with a discussion of socio-economic systems? For reasons that will become clearer as we go along, I shall again choose the latter, and, without further preamble, now broach the question of the adaptive properties of the two great socio-economic systems that influence human behavior in our time: capitalism and socialism.

Our choice of approach requires us to begin with the seemingly simple, but actually very difficult, task of making clear what we mean by "capitalism" and "socialism." If we begin with capitalism, I do not think there will be much disagreement as to the necessary elements that must go into our basic definition. Capitalism is an *economic* order marked by the private ownership of the means of production vested in a minority class called "capitalists," and by a market system that determines the incomes and distributes the outputs arising from its productive activity. It is a *social* order characterized by a "bourgeois" culture, among whose manifold aspects the drive for wealth is the most important.

As we shall see, this deceptively simple definition has unexpectedly complex analytical possibilities. But it also calls our attention to the necessity of conducting our inquiry at a suitable level of abstraction. It is the behavior of general socio-economic *systems* in which we are interested, not the behavior of particular examples of those systems. This is a consideration that has special relevance for the political animus that we carry with us in an investigation of this sort. It is a common tendency, for example, for radical analysts to assume that the word "capitalism" is synonymous with the words "United States." "The United States is a capitalist society, the purest capitalist society that ever existed," according to Paul Sweezy, the foremost American Marxian critic. Similarly, the French Marxist Roger Garaudy has agreed, "The capitalist system in its most typical, richest, and most powerful expression, is that of the United States. . . ."[1]

Serious problems arise from the choice of the United States, not as the richest or most powerful, but as the *typical* capitalist nation. The first is the assumption that certain contemporary attributes of the United States (racism, militarism, imperialism, social neglect) are endemic to all capitalist nations—an assumption that opens the question of why so many of these features are not to be found in like degree in all capitalist nations (for instance, England or Sweden or the Netherlands), as well as why so many of them are also discoverable in non-capitalist nations such as the Soviet Union.

Second, the selection of the United States as the archetype of capitalism raises awkward issues with regard to socialism. For the logical question then is: If the United States is chosen to represent "typical" capitalism by virtue of its size, power, or global predominance, must we not designate the Soviet Union as the "typical" socialist nation for the same reasons?

1. Paul M. Sweezy, "The American Ruling Class," in *The Present as History* (Monthly Review Press, 1953), p. 126; Roger Garaudy, *Marxism in the Twentieth Century* (Scribner, 1970), p. 13.

The radical critic recoils at this "logic," and explains the repugnant features of Soviet Russia as the unhappy legacy of its past, a tragic instance of the socialist ideal fatally compromised by the institutional and historical setting in which it was first achieved. But if we take this argument to be valid—and surely it has serious claim to consideration—are we not forced to extend the same apologia to the United States? That is, does not the United States then appear, not as a "pure" realization of capitalism, but as a deformed variant, the product of special influences of continental isolation, vast wealth, an eighteenth-century structure of government, and the terrible presence of its inheritance of slavery—the last certainly not a "capitalist" institution? Indeed, could we not argue that "pure" capitalism would be best exemplified by the economic, political, and social institutions of nations such as Denmark or Norway or New Zealand?

The point of this caution, which applies equally to the conservative who singles out the Soviet Union as the incarnation of socialism, is that we cannot analyze the adaptive properties of capitalism or socialism by confining our attention to the merits or shortcomings of any single example of either system. The range of social structures, traditions, institutions of government, and variations of economic forms is so great for both socio-economic orders that generalizations must be made at a very high level of abstraction—so high, in fact, that one may seriously question whether an analysis along these lines can shed much light on the adaptive capabilities of, say, "capitalist" Sweden or Japan versus "socialist" Hungary or East Germany.

Why, then, pursue at all the elusive question of the capabilities of these socio-economic orders? Two reasons seem cogent. First, the words "socialist" and "capitalist" continually recur in day-to-day (or in scholarly) discussions of the future, and therefore it seems worthwhile to examine the specificity that can be given to these terms, even if it turns out to be very small. Second, I believe a socio-economic analysis is warranted because, for all the variety in national forms, both systems must cope with common problems rooted in their economic and social underpinnings. That their responses may differ widely does not lessen the importance of singling out these common problems and examining the challenges they present to the family of related societies in which they appear.

Can we make a plausible prognosis with regard to capitalism as an "ideal type"?

Our first answer must be a disappointing one. On the basis of the bare specifications of capitalism two major historic projections for that system have been constructed—one negative, one positive—both of which have been demonstrated to be inadequate. The negative prognosis is most forcefully conveyed by the Marxian "scenario" for capitalist development, a scenario foretelling its gradual polarization into two bitterly inimical camps, its growing inability to maintain a smoothly functioning economic process, and its eventual collapse through revolution. Central to that prophecy was the expectation that the dynamics of the system would create a working class "ever increasing in numbers," and disciplined by its economic hardships into an instrument of revolutionary historic change.

Some of that prediction, it should be noted, has been validated. The dynamics of capitalism did bring about a steady forced migration of farmers and self-employed small proprietors into the ranks of wage and salary workers, and the pronounced instability of the system did generate recurrent severe economic hardships. What seems to have forestalled the final vindication of the Marxian prognosis, however, was a series of developments that offset the revolutionary potentialities envisaged by its author. One such offsetting tendency was the steady augmentation of per capita output, which effectively undercut the development of proletarian feelings of exploitation. A related development was the rise of a "welfare" framework that also served to defuse the revolutionary animus of the lower classes. Last and perhaps most important was the gradual discovery—a discovery both in economic techniques and in social viewpoint—that government intervention could be used to prevent a recurrence of the near-catastrophic collapses suffered by the laissez-faire versions of capitalism characteristic

83

of the late nineteenth and early twentieth centuries.

As we shall see, the Marxian conception of capitalism as a system inherently burdened with internal "contradictions" is far from being disproved by these events. But, meanwhile, what of the positive prognostication with regard to capitalism? Unlike the radical scenario, the second prognosis has had no single major expositor. It is to be found, rather, in the generally shared expectations of such writers as Alfred Marshall and John Maynard Keynes, or indeed the main body of non-radical twentieth-century economists.

Their prediction, like that of the Marxists, was also based on the presumed behavior of a private-property, market-directed, profit-seeking system, but not surprisingly it emphasized elements that were overlooked in the radical critique. The basic prognosis of the conservatives was that the capitalist system would display a steady tendency to economic growth, and that the socially harmful results of its operations —poverty, social neglect, even unemployment—could be effectively dealt with by government intervention within the institutional framework of private property and the market. As a result, the conservative view projected a trajectory for capitalism that promised the exact opposite of the Marxian: economic success coupled with a rising degree of social well-being.

Yet that prediction has also not fully materialized. As with the Marxian prophecy, certain of its elements were in fact attained, in particular an increase in per capita output and an expansion of social welfare policies. But the social harmony that was expected to result from these trends did not follow along. In the United States, for example, the economic transformation from the depressed conditions of the 1930s to those of the 1970s—a transformation that effectively doubled the real per capita income of the nation —failed to head off racial disturbances, an explosion of juvenile disorders among the affluent as well as among the poor, a widespread decay in city life, and a serious deterioration in national morale. And this disturbing experience has not been confined to the United States. Unprecedented economic growth in France and Germany and Japan has not prevented violent outbreaks of disaffection in those countries, especially among the young. Nor have Sweden and England and the Netherlands—all countries in which real living standards have vastly improved and in which special efforts have been made to reduce the economic and social distance between classes—been spared similar expressions of an underlying social discontent.

This failure of social harmony to accompany economic growth was as fundamentally disconfirming for the validity of the conservative prognosis as was the failure of a revolutionary temper to emerge for the radical prognosis. What explanation can we give? We can only hazard a few guesses. One is that poverty is a relative and not an absolute condition, so that despite growth, a feeling of disprivilege remains to breed its disruptive consequences.[2] Another is that each generation takes for granted the standard of living that it inherits, and feels no gratitude to the past.[3] Finally, the failure of the conservative prognosis may simply signal the possibility that whatever its economic strengths, the social ethos of capitalism is ultimately unsatisfying for the individual and unstable for the community. The stress on personal achievement, the relentless pressure for advancement, the acquisitive drive that is touted as the Good Life—all this may be, in the end, the critical weakness of capitalist society, although providing so much of the motor force of its economy.

The lesson of the past may then only confirm what both radicals and conservatives have often said but have not always really believed—that man does not live by bread alone. Affluence does not buy morale, a sense of community, even a quiescent conformity. Instead, it may only permit larger numbers of people to express their existential unhappiness because they are no longer crushed by the burdens of the economic struggle.

Does this confounding of two contrary prognoses leave us with anything on which to base a general estimate of capitalism as a system capable of meeting

2. See Richard Easterlin, "Does Money Buy Happiness?" *The Public Interest*, Winter 1973.
3. See Paolo Leon, *Structural Change and Growth in Capitalism* (Johns Hopkins Press, 1967), pp. 23f.

the problems of the future? We will be able to answer that question better after we have looked at the other side of the coin, and applied to socialism the same "ideal-typical" scrutiny that we have so far applied only to capitalism.

Here we must begin by recognizing a serious difficulty. In discussing capitalism as an ideal type, we had in mind a variety of "advanced" nation-states that, however different in many aspects, all shared a roughly similar social setting. Tacitly our analysis of capitalism referred to a group of nations characterized by a common "bourgeois" style of life and by highly developed industrial structures with their associated common aspects of mass production and high-level consumption.

When we turn to the consideration of socialism, no such unified image presents itself. We can easily describe socialism as an *economic system* by its replacement of private property and the market with some form of public ownership and planning. But socialism is much more difficult to specify as a social order than is capitalism. Indeed, we can identify at least two, and possibly three, social orders that rest on public property and planned economic activity.

One of these is typified by the industrial "socialism" of present-day Russia and much of Eastern Europe. Characteristic of this type of socialism are two salient features: an industrial apparatus closely resembling that of capitalism, both in structure and in outlook, and a highly centralized, bureaucratic, and repressive social and political "superstructure." A second "socialist" order is represented by the societies that have arisen in the underdeveloped world, or that are likely to emerge there in the future. Here political centralization and social repression exist, but not the framework of industrialism characteristic of the first type. This is the obvious consequence of present-day underdevelopment itself, but we must also consider the possibility that these nations will seek future development along lines that minimize such an industrial structure.

A third type of socialism presents far more difficulties for our kind of analysis than the other two, because it exists partly in historical fact, partly in imagination. This is a socialist order that seeks to combine a high degree of industrialism with a considerable amount of political freedom and decentralization of control. This form of socialism has been perhaps most closely approximated in the brief tragic career of "socialism with a human face" in Czechoslovakia and—to an extent difficult to determine—in contemporary Yugoslavia. More important, perhaps, is its existence in the minds of many socialist reformers as the kind of society toward which Western socialism may hope to move in the foreseeable future. It therefore exerts its influence as a historical force, even though its realization in fact is as yet very slight.

It will be necessary, therefore, to proceed with

great caution in attempting to describe the dynamics of the family of socialist societies. Nonetheless we can at least start with a striking fact. It is that the two main prognoses with respect to *industrialized* socialism have been proved as inadequate as did the corresponding prognoses with respect to industrial capitalism.

The first of these prognoses, frequently encountered only a generation ago, was that industrial socialism was "impossible," and that socialist economies would break down by virtue of their inherent irrationality. The resemblance of this prediction to that of the Marxian expectations with regard to the malfunction of capitalism is evident, and so is the failure of the prediction to come true. Despite the inability of industrial socialist economies to work with the smooth efficiency expected by their partisans—indeed, despite the frequent vindication of their critics' expectations of irrationality and malperformance—socialism did not break down. If economic discontent here and there reached threatening levels, the same can be said for the capitalist world in the 1930s. But in the one case as in the other, tendencies to growth overcame those of stagnation or crisis, so that by strictly economic criteria, industrial socialism proved as great a "success" as did capitalism.

But with the disappearance of the once confidently advanced prediction of a spontaneous collapse of socialist economies there also came a fading of a second prognosis, the counterpart to the sanguine out-

89

look for capitalism. This was the belief that the replacement of private ownership by public ownership, and the displacement of the market by planning, would usher in an age of high social morale as well as high economic performance.

Again in striking parallel to the disappointments that have attended the growth of economic output under capitalism, the "successful" workings of socialist economic institutions have not brought the hoped-for results. On the contrary, if we are to judge by the relentless campaigns in the Soviet or East European press against absenteeism, carelessness, bureaucratic tyranny, or "un-socialist" attitudes, or by the actual revolts of workers in Poland and Hungary against their working conditions, or by the widespread evidence of a sense of intellectual oppression in many socialist nations, the social results of socialist economic growth have been very disappointing. If some of the more extreme forms of social disorder characteristic of the West, above all the anti-establishment mood and actions of youth, are much less to be observed, there seems good reason to credit this to the efficiency of the socialist police rather than to an absence of such tendencies on the part of the young. As evidence that these socialist nations have not generally attained their hoped-for level of communal spirit, there is the damning fact of the continuous efforts of their citizens at all levels of society to emigrate to capitalist nations, and the equally damning refusal of their authorities to permit the free entry of ideas.

These generalizations must, however, be more discriminatingly applied to "socialism" in general, with its range of types, than to the more unified cultures of bourgeois societies. In the underdeveloped nations, above all in China and Cuba, extraordinary efforts have been made to generate a high sense of morale, certainly with some success. We shall therefore look into the case of "under-developed" socialism more carefully in our next chapter. In Czechoslovakian or Yugoslavian socialism, the negative findings must also be tempered, although it is difficult to know by how much. But I think the evidence, even including these latter nations, is sufficient to enable us to assert that the economic success of industrial socialism, in and of itself, has not brought a corresponding rise in general "happiness" or social contentment, much as with the mixed record of economic success and social disappointment of capitalism.

I do not make this assertion to claim that industrial socialism has therefore failed: on the contrary, I imagine that in the minds of the majority of its citizens it has "succeeded," to much the same degree as capitalism. Rather, I call attention to the situation within the industrial socialist world to stress the surprising similarity of outcomes between two otherwise widely differing systems. Each has been marked with serious operational difficulties; each has overcome these difficulties with economic growth. Each has succeeded in raising its level of material consumption; each has been unable to produce a climate of social satisfaction.

91

This leads to the suggestion that common elements of great importance affect the adaptability of both systems to the challenges of the human prospect.

In the light of our analysis in the preceding chapter, it will not come as a surprise if I identify these common elements as the forces and structures of scientific technology on which both systems depend for their momentum. This suggestion would least seem to need supporting argument in explaining the ability of both systems to achieve economic growth, despite the malfunctions of the market in one case and of planning machinery in the other. All the processes of industrial production that are the material end products of scientific technology have one characteristic of overwhelming effect—their capability of enormously magnifying human productivity by endowing men with literally superhuman abilities to control the physical and chemical attributes of nature. Once an industrial system has been established—a historic process that has been as painful for capitalism as for socialism—it truly resembles a gigantic machine that asserts its productive powers despite the sabotage of businessmen or bureaucrats.

It is perhaps less self-evident that the common disappointments of capitalism and socialism with regard to the achievement of "happiness" can also be traced to the presence of scientific technology and the industrial civiliation that is built upon it. I have already pointed out the peculiar ills that may have their roots in the capitalist ethos, and it is also clear that many of the socialist dissatisfactions arise from the repressive political and social institutions to which we have called attention. Nevertheless, if we look more deeply I think we can find a substratum of common problems that spring from the industrial civilization of both systems.

For industrial civilization achieves its economic success by imposing common values on both its capitalist and socialist variants. There is the value of the self-evident importance of efficiency, with its tendency to subordinate the optimum human scale of things to the optimum technical scale. There is the value of the need to "tame" the environment, with its consequence of an unthinking pillage of nature. There is the value of the priority of production itself, visible in the care both systems lavish on technical virtuosity and the indifference with which both look upon the aesthetic aspects of life. All these values manifest themselves throughout bourgeois and "socialist" styles of life, both lived by the clock, organized by the factory or office, obsessed with material achievements, attuned to highly quantitative modes of thought—in a word, by styles of life that, in contrast with non-industrial civilizations, seem dazzlingly rich in every dimension except that of the cultivation of the human person. The malaise that I believe flickers within our consciousness thus seems to afflict industrial socialist as well as capitalist societies, because it is a malady ultimately rooted in the "imperatives" of a common mode of production.

I am aware, of course, that it is questionable to

assert that technology has "imperatives," for technology is no more than a tool in the hands of man. If the industrial apparatus has imposed its dehumanizing influence on capitalist and socialist industrial societies alike, there remains the possibility that in another milieu that apparatus could be turned to human account. It may be that extensive decentralization, workers' control, and an atmosphere of political and social freedom could better reconcile the industrial system with individual contentment.

I will not hide my doubts, however, that these reforms can wholly undo the de-humanizing requirements of an industrial system. Modes of production establish constraints with which humanity must come to terms, and the constraints of the industrial mode are peculiarly demanding. The rhythms of industrial production are not those of nature, nor are its necessary uniformities easily adapted to the varieties of human nature. While surely capable of being used for more humane purposes than we have seen hitherto, while no doubt capable of greater flexibility and much greater individual control, industrial production nonetheless confronts men with machines that embody "imperatives" if they are to be used at all, and these imperatives lead easily to the organization of work, of life, even of thought, in ways that accommodate men to machines rather than the much more difficult alternative.

The suggestion that a common industrial organization of life is responsible for certain parallels in the development of capitalism and industrial socialism can be no more than a speculation. More pressing are the immediate challenges that both great socio-economic orders will have to face.

We have already seen that the problem of population growth must be discussed in terms of the differential rates of growth of the developed and the underdeveloped lands. The question to be considered, then, is whether the dangerous consequences of the population problem in the underdeveloped world will, in the end, affect industrial socialist nations, such as the Soviet Union or East Germany, differently from capitalist nations, such as the United States or West Germany. These consequences, we will recall, resided in the encouragement given to the emergence of revolutionary regimes, and in the temptation—or necessity— for these regimes to use nuclear blackmail as a means of inducing the developed world to transfer its wealth on an unprecendented scale to the underdeveloped regions.

In this impending drama, it seems likely that the advanced socialist world will be the initial beneficiary of feelings of comradeship from the new revolutionary nations, and will probably be their immediate benefactor as well. Conversely, the rise of revolutionary governments presents the danger that capitalist nations will be tempted to use force to keep the spread of revolutionary socialism within bounds. The Indochina war, among whose motives the "containment" of communism was certainly a major element, is an all too clear example of precisely this form of counter-revolutionary activity. Thus, the population problem brings as an immediate consequence an aggravated risk of aggressive behavior on the part of the threatened capitalist world.

These reflections apply, however, mainly to the short run, when the setting of international existence will be much as we find it today. In the longer run the prospect alters considerably. To begin with, over a longer span we must resist the temptation to generalize from United States' belligerence in the past decade as firmly as we must resist similar generalizations based solely on the behavior of the Soviet Union. Looking over the record of capitalist nations during the past century, one does not discover a universal tendency toward military activity. The pacific attitude of the Scandinavian bloc, or of the smaller countries of Europe, the anti-military record of the United States until World War II (despite its punctuation by limited imperialist adventures), the recent disappearance of traditional warlike attitudes from the cockpit of capitalist conflict in Europe, make one cautious in declaring that capitalism is "inherently" a war-prone system. Moreover, in examining the motives that provoked the major capitalist wars during that century, one discovers, in addition to the specifically capitalist drives for economic expansion, powerful considerations of national prestige, strategic geographic advantage, or simply ideological enmity—all motives that have driven nations long before the advent of capitalism as a system and that continue to manifest themselves visibly in the behavior of socialist nation-states.

More important yet, there is reason to believe that the pressures of the population explosion will come to bear increasingly on all nations alike, socialist as well as capitalist. The initial congruence of political interests between young revolutionary regimes and older industrial socialist ones must contend with a growing conflict over their economic aims. Given the closing vise of resource and energy supplies, and the gradually approaching barriers to growth imposed by the environment, it is clear that control over the planet's resources and claims on its output must become problems that will increasingly threaten the viability of all industrial systems. In the inescapable competition for dwindling resources and for the right to maintain, if not increase, the level of national output, I can see no reason why the imperatives of self-preservation should not operate as strongly among the socialist industrial nations as among capitalist ones. In both cases, wars of "preemptive seizure" would be a possible strategy. Barring such undertakings, I do not see why demands for a more equitable sharing of the world's output should not be as peremptorily directed by the poor countries against their rich socialist brothers as against their rich capitalist enemies.

The long-run problem then will be that of coping

97

with a "two-level" world. Whether or not that problem can be resolved without recourse to war, initiated by the poor countries or by the rich ones, cannot be foreseen. Much hinges on the degree of reason, compassion, or flexibility with which one endows the imaginary capitalist and socialist nations or ruling classes of the future, a matter in which our political presuppositions strongly affect our judgments.

This estimate need not be wholly subjective, however. For it becomes increasingly clear that the central issue of the future will lodge in the capability of dealing with the environmental limitations that emerge ever more insistently as the most intransigent of the problems of the future. Let us therefore ask what can be said as to the relative abilities of capitalist and socialist systems in coping with that challenge.

To start with the capitalist side, there is no doubt that the threatened depletion of resources, and the drastic ecological dangers that loom at a somewhat greater distance, directly threaten a main characteristic of capitalism—its strong tendency to expand output. This tendency serves three main functions for the system. It expresses the drives and social values of its dominant class. It provides the means by which a market-coordinated system can avoid the dangers of a "general glut." And, finally, it accommodates the striving of its constituents for larger rewards. Thus, expansion has always been considered as inseparable from capitalism, whether as a necessary condition for its operation, as Marxian critics would claim, or as a justification for the institutions of private property and the market, as the conservative protagonists of capitalism have maintained. Conversely, a "stationary," non-expanding capitalism has always been considered either as a prelude to its collapse or as a betrayal of its historic purpose.

Is a stationary capitalism therefore unworkable? Is it a contradiction in terms? The answers are not open and shut, for they depend on various sociological assumptions from which, I need hardly add, our subjective evaluations cannot be wholly excluded. To begin with the first of the functions served by expansion, I do not think that one can make a dogmatic assertion that the social values and drives of its dominant class could not be accommodated within a largely static framework. Here we have the evidence of the extremely defensive economic posture characteristic of French or English capitalists just before and after World War II, respectively, and of the curiously bureaucratic complexion of Japanese capitalism, run by an extraordinarily "passive" and conformist managerial elite.

The expansive drive of capitalism springs, however, not only from the "animal spirits" of its dominant class but also from the restless self-aggrandizing pressures of its corporations. Here as well, however, a solution is imaginable. Much of the aggressive drive of firms arises from their continuous striving for a larger share of the market. A deceleration in growth, enforced by government decree, could include provisions for leaving market shares relatively undisturbed. Such a solution would be no more than a full-fledged transformation of "private" capitalism into planned "state" capitalism—a transformation already partially realized in Japanese capitalism.

It is perhaps less simple to construct a plausible stationary capitalism in the light of the second function served by expansion; namely, the avoidance of a severe economic crisis. As economists from Adam Smith and Marx through Keynes have pointed out, a "stationary" capitalism is subject to a falling rate of profit as the investment opportunities of the system are used up. Hence, in the absence of an expansionary frontier, the investment drive slows down and a deflationary spiral of incomes and employment begins.

Yet I do not think that one can maintain that a stationary capitalism is therefore "impossible." Expansion serves an indispensable purpose in maintaining a socially acceptable level of employment and demand in laissez-faire capitalism. It is by no means so certain that it is indispensable in a managed state capitalism. There seems no inherent reason why the deflationary tendencies of such a system could not be offset by a variety of measures. A high level of public demand could be provided by government investment in housing, education, and the like, by transfer payments within the nation, or by the distribution of "surplus" goods, if any existed, to the underdeveloped nations. All these measures are already in use in various parts of the capitalist world. Thus a stationary state would not seem to present insuperable problems for a managed capitalism, insofar as those problems concerned the maintenance of employment or aggregate purchasing power.

It may be argued that I have leaned over much too far in projecting such an optimistic prognosis for capitalist adaptation to a non-expansionary situation. I have done so deliberately, however, because there remains an aspect of the transition to a stationary system that strikes me as far more taxing with respect to capitalist powers of accommodation. This is the problem of finding a means of managing the social tensions in a capitalist system in which growth had ceased or was very greatly reduced.

Central to capitalism, as we have already noted, is a bourgeois ethos of economic advancement. Previously we have suggested that this ethos may be partly responsible for the failure of capitalist expansion to produce high social morale. But the pervasive values of competitive striving and expected personal advancement also present another problem—how to satisfy the demands of the lower and middle classes for higher living standards, while protecting the privileges of the upper groups. The solution has been to increase the output of the economy, thereby providing absolute

101

increases in income to all classes, while leaving the relative share of the upper groups relatively undisturbed.

The prospect of a stationary economy directly challenges this traditional solution. For under a stationary (or even a slow-growing) capitalism, continued efforts of the lower and middle classes to improve their positions can be met only by diminishing the absolute incomes of the upper echelons of society. A stationary capitalism is thus forced to confront the explosive issue of income distribution in a way that an expanding capitalism is spared.

In this connection we must bear in mind that we are not merely talking about the dismantling of a few vast fortunes or the curtailment of a handful of swollen incomes, although that might be difficult enough. What is at stake are the incomes of the upper middle classes, which include something like the upper fourth or fifth of the nation. This upper stratum is by no means composed of millionaires alone, but also includes teachers, shopkeepers, professional and technical workers: in the United States in the early 1970s a family entered the upper fifth with an income of about $15,000.

This stratum of society enjoys about 40 percent of the nation's total income. If the pressure from below were to eliminate its advantage over the "average" family, the upper stratum would have to yield a large fraction of its income, from perhaps a third at its lower levels to well over half at its upper levels. This gives one some appreciation of the magnitude of the political strain to which a massive pressure for income redistribution would give rise.

One saving possibility must, however, be considered in assessing this strain. It is possible that growth could be permitted to continue for an indefinite period, provided that it were confined to outputs that consumed few resources and generated little heat. An expansion in the services of government, in the administration of justice, in the provision of better health and education, arts and entertainment, would not only rescue the system from a fatal encounter with the environment but might produce enough "growth" to ease the income distribution problem.

If capitalism is to survive for a considerable period, this is the road along which it will assuredly have to travel. Perhaps in some cases it may successfully manage such a shift in the composition of its output. But we must not lose sight of the global setting in which this shift must be made. A transition to a more equitable distribution of income within the capitalist nations will have to take place at a time when the larger struggle will focus on the distribution of resources among nations. If this struggle is gradually decided in favor of the underdeveloped world, whether out of humanitarian motives, the pressure of nuclear blackmail, or simply by the increased political cohesion and bargaining power of the poorer regions, the citizenries of the wealthy nations will find themselves in a long period of declining physical output per capita. This is apt to be the case, even without an international redistribution, if the many constraints of the environment exert their expected effects, beginning perhaps as soon as the coming decade.

Thus the difficulty of managing a socially acceptable distribution of income in the capitalist nations is that it will have to contend with the prospect of a decline in the per capita output of material goods. The problem is therefore not merely a question of calling a halt to the increasing production of cars, dishwashers, or homes while encouraging the output of doctor's services or the theater, but of distributing a shrinking production of cars, appliances, homes. The difficulties of a limited oil shortage have brought home to many Americans the hitherto unimaginable possibility that their way of life might not be indefinitely sustainable. If that shortage is extended over the next generation or two to many kinds of material outputs, a climate of extreme "goods hunger" seems likely to result. In such a climate, a large-scale reorganization of social shares would have to take place in the worst possible atmosphere, as each person sought to protect his place in a contracting economic world.

I am inclined to the belief, therefore, that the problem of income distribution would pose extreme difficulties for capitalism of a political as well as an economic kind. The struggle for relative position would not only pit one class against another, but also each against all, as lower and middle groups engaged in a free-for-all for higher incomes. This would bring enormous inflationary pressures of the kind that capitalism is already beginning to experience, and would require the imposition of much stronger control measures that any that capitalism has yet succeeded in introducing—indeed, than any that capitalist governments have yet imagined.

In bluntest terms, the question is whether the Hobbesian struggle that is likely to arise in such a strait-jacketed economic society would not impose intolerable strains on the representative democratic political apparatus that has been historically associated with capitalist societies.

It is of course, foolish to suggest that capitalism is the *sine qua non* of democracy, or to claim that democracy, with its commitment to political equality, does not conflict in many ways with the inequalities built into capitalism. Nonetheless, it is the plain historic fact that bourgeois societies have so far succeeded to a greater degree than any other social order in establishing parliamentary procedures, independent judiciaries, and constitutionally limited executives, all essential elements in a democratic political system. The question to be faced, then, is whether these political institutions can be expected to cope with the social and economic transformations whose extensive character we have indicated.

Here prediction along the lines of an "ideal type"

cannot bring us very far. It is possible that some capitalist nations, gifted with unusual political leadership and a responsive public, may make the necessary structural changes without surrendering their democratic achievements. At best, our inquiry establishes the approach of certain kinds of challenges, but cannot pretend to judge how individual nations may meet these challenges. For the majority of capitalist nations, however, I do not see how one can avoid the conclusion that the required transformation will be likely to exceed the capabilities of representative democracy. The disappointing failures of capitalist societies to create atmospheres of social harmony, even in expansive settings, does not bode well for their ability to foster far-reaching reorganizations of their economic structures and painful diminutions of privilege for their more prosperous citizens. The likelihood that there are obdurate limits to the reformist reach of democratic institutions within the class-bound body of capitalist society leads us to expect that the governments of these societies, faced with extreme internal strife or with potentially disastrous social polarization, would resort to authoritarian measures. To the extent that these measures would necessarily include the national management of corporations and the non-market determination of income levels, the direction of change might be described as a movement toward "socialism," although in a manner very different from that of the classic revolutionary scenario and with implications that will distress the partisans of socialism as a democratic form of government.

These reflections raise a question that may have been impatiently waiting in the reader's mind. After all, the ecological threat is still some distance in the future. Hence long-term speculations as to the feasibility of a stationary capitalism may seem hopelessly academic in the face of nearer-terms risks of war, or of the disruption of capitalism from other causes, such as its inability to generate a high enough social discipline and morale. That may indeed be the case. But if capitalism collapses, what next?

As we have already seen, the successor may well be a severe authoritarian regime that is not easy to analyze in terms of our socio-economic ideal types. But let us suppose that the collapse of capitalism would usher in socialism—that is, a society built on the public ownership of goods and the replacement of the market by widespread planning. What can we say about the capabilities of such a system before the demands of the environmental challenge?

Here the possibilities of applying a socio-economic analysis seem much simpler. It appears logical to conclude that socialism, with its direct commitment to a planned economy and with its freedom from the ideological blockages of private property, could manage the adaptation of an industrial society to a stationary equilibrium much more readily than capitalism.

I believe this is true in the short run. Over a longer period, however, grave problems would emerge. A socialist society would also have to achieve a politically acceptable distribution of its income among its people. The task of arriving at such as division of income would be much more difficult in a period of shrinking physical output than in an economy where all levels expected their real incomes to rise. Hence a democratically governed socialism would very likely face the same Hobbesian struggle for goods as a democratically governed capitalism; and whereas an authoritarian socialism could certainly enforce some kind of solution, it seems likely that this would entail a degree of coercion that would make "socialism" virtually indistinguishable from an authoritarian "capitalism."

The similarity of the problems of and responses to the stationary state for both socialism and capitalism brings us finally to confront a question that has persisted like a *basso ostinato* through these pages. This is the relation of the two systems to the industrial civilization that has again and again emerged as a root cause for the dangers of the human prospect and as the common basis for the economic successes (and perhaps the social failures) of capitalism and industrial socialism. Is it now possible to maintain, on the grounds of our socio-economic analysis, that socialism will have a significant advantage over capitalism in asserting the necessary controls over the runaway forces of science and technology?

Once more I believe we must differentiate between the short-run and long-run capacities for response. In the short run, as in the case of international tensions and in the initial stages of coping with the pressures of a stationary economy, I would think that industrial socialism would possess important advantages. The control over the direction of science, over its rate of incorporation in technology, and over the pace of industrial production as a whole should be much more easily achieved in a society that does not have to deal with the profit drive, than in one that does. To be sure, socialist systems have their own handicaps in the bureaucratic inertias of planning. But the absence of a necessity to heed the pull of commercial considerations should nonetheless confer an additional degree of social flexibility to the socialist control over the industrial process.

In the longer run, however, I believe that we find a convergence of problems, as we have so often before. For what portends, in that longer run, is a challenge of equal magnitude for industrial socialism as for capitalism—the challenge of drastically curtailing, perhaps even dismantling, the mode of production that has been the most cherished achievement of both systems. Moreover, that mode of production must be abandoned in a mere flash of time, as historic sequences are measured. Given the present pace of industrial growth—which will take prodigies of science to maintain in the face of dwindling resources—the edge of the heat emission danger zone may be reached in as little as three or four generations. Failing the achievement of the needed scientific breakthroughs, we will be spared the heat barrier simply because we will be unable to produce the energy or to process the resources to maintain our present growth rates.

Thus, *whether we are unable to sustain growth or unable to tolerate it*, there can be no doubt that a radically different future beckons. In either eventuality it seems beyond dispute that the present orientation of society must change. In place of the long-established encouragement of industrial production must come its careful restriction and long-term diminution within society. In place of prodigalities of consumption must come new frugal attitudes. In these and other ways, the "post-industrial" society of the future is apt to be as different from present-day industrial society as the latter was from its pre-industrial precursor.

Can we expect an industrial socialist society, be it characterized by authoritarian or by democratic government, to weather such a transformation more easily than a capitalist society, "private" or state? I doubt it. Both socio-economic systems are committed to a civilization whose most striking aspect is its productive virtuosity. But my skepticism is based on more than the resistances and inertias of vested interests that we find throughout history when established modes of production become obsolete. It is also founded on a political consideration, namely, whether *any* society can bring about alterations of this magnitude through the conscious intervention of men, rather than by convulsive changes forced upon men. I cannot hope to substantiate this judgment until we have looked into the political and psychological dimensions of the human capacity for response. This is the aspect of the human prospect to which we must therefore now turn our attention.

AFTERWORD

THE PRECEDING CHAPTER mulls over the long-term adaptive capabilities of capitalism and socialism. It is necessarily conjectural and uncertain, but I cannot improve it; and I will let it stand. But there is something I must add about the short-term problems of capitalism that was not visible when I weighed things in the balance in 1972. This is the shared experience of economic crisis I wrote about in reconsidering Chapter I.

Capitalism has been in a state of crisis for the last several years, a crisis that began with the OPEC "oil shock" to which we have already referred. Oil shock hit the industrial world like an immense economic earthquake. It brought to an almost instant halt the long boom that had buoyed the capitalist world since the early 1950s—the longest uninterrupted boom in its

111

history. Within a few months production began to fall, and unemployment to rise, around the capitalist world. In the United States, production declined by about 9 percent and unemployment rose by 85 percent—far less severe in both cases than the trauma of the Great Depression, but a sign of serious economic disruption, nonetheless. And with minor variations, this pattern of malfunction was repeated from Australia to Austria, South Africa to Sweden.

It was not just the downturn in growth or the rise in unemployment that signaled the advent of the crisis. It was the perplexing and worrisome fact that despite the fall in output and the rise in joblessness, price levels continued an upward climb that was already a twenty-year-old problem. Indeed, the pace of inflation actually accelerated with oil shock. In the five years before the 1973 OPEC price rise, the cost of living in the United States had been rising at about 5 percent a year. Beginning in that year, the rate of annual increase jumped to six, then to nine, then to double digit rates before falling back somewhat. A similar pattern appeared in other nations. In France prices rose by 75 percent in the five years after oil shock, in Italy by 125 percent, in England by 185 percent—in all cases accelerating sharply in contrast to pre-1973 rates.

Of course the rising price of oil was itself responsible for some of the increase in inflation. But in general, the increased cost of fuel accounted for no more than a tenth to a quarter of the jump in price levels.

112

The rest was a reaction to oil shock, as if the disturbance to the economy exacerbated tendencies that were already present and that only needed an external blow to become more pronounced.

What are these latent tendencies? Economists are by no means of one mind as to the answer. But I suspect this is the case because they tend to regard inflation as an ailment inflicted on the economy from without, by oil shock or government spending or excessive wage hikes, rather than attempting to understand inflation as a manifestation of a condition generated by capitalism from within. *That condition is the instability that has always been the hallmark of capitalism, the consequence of what Karl Marx called its "anarchy"—its uncoordinated thrusting expansiveness.*

In retrospect, we can see the cumulative depressions of the past as the way in which that instability manifested itself in an era when capitalism had not yet developed the defenses of government spending and social insurance. What we have yet to see is that modern inflation can be viewed as the obverse of depression—that is, as the manner in which instability appears in a capitalist system that has altered its institutions to guard against massive depression, but in so doing has created conditions in which prices have an upward impetus. For it is clear that some such shift in the direction of instability has taken place. Gone is the tendency for recession to snowball into massive

113

depression. During the OPEC crisis, unemployment rose by 85 percent, not by 700 percent as it did in the 1930s. Production fell by 9 percent, not by 50 percent. The difference is what happened to prices. In the 1930s they fell. In the 1970s, they exploded upward.

What causes instability to vent itself in inflation rather than depression? At bottom, the reasons lie in institutional changes that profoundly differentiate present-day capitalism from the capitalism of an earlier time. One such difference is that unions are much more powerful today, especially in the industrial core. A second difference is that prices in many industrial markets are nowadays "administered" by powerful corporations rather than opened to the cross currents of competition. And the most significant difference of all is the vastly larger presence of government, not only as an independent economic force on its own account, but as the provider of social support and security to millions.

All these institutional changes have exerted their influences in a common direction. They inhibit the unrestrained downward tendencies of depression by bolstering public expenditures and by providing a cushion of economic support to households. As a result, output declines only to a limited degree. But at the same time, the institutional changes encourage the upward expression of instability by increasing the pressure on costs and prices. Whether or not output falls, stronger unions win higher wage settlements, and make wage reductions next to impossible. Oligopolistic companies acquiesce in these wage settlements because they know they can recoup their costs by raising their prices as a group. The larger government sector directly pushes prices upwards through its purchasing activities, and indirectly pushes them up by providing financial support to millions of households.

Thus the altered institutional setting changes the perennial tug of war within capitalism in favor of higher prices and costs, and once this alteration becomes recognized as permanent, it begins to perpetuate—indeed, to exacerbate—itself. Labor unions, anticipating higher prices, seek wage settlements that will put their members ahead of the rising cost of living. Corporations write escalator provisions for rising prices and costs into their contracts with suppliers and customers. Households, distrusting the government's ability to restrain inflation, develop a low grade panic that leads to buy now, pay later. Thus the very recognition of an endemic state of inflation begins to set into motion the Hobbesian free-for-all we spoke of in our chapter as a prospect for capitalism at a later state of tension.

This is not, of course, a full explanation of inflation. Other elements, including the impact of domestic monetary policy and foreign international policy, must also be taken into account, as well as the effects of crop failures, resource squeezes, productivity lags, and the like. But the essential conclusion is clear enough. It is that inflation (coupled with mild recession) is best viewed as the manner in which the endemic instability of capitalism is manifested in an era of big labor, oligopolistic business, and interventionist-minded government, just as severe depression was the way in which the instability was evidenced in a period of weak labor, more competitive business, and laissez-faire government.

Moreover, viewing inflation in this way enables us to see that the adaptation to inflation and depression are much more closely linked than at first appears. For when we look back to the 1930s, we discover that many economists and statesmen knew how the Great Depression could be cured. It would require enough government expenditure to offset the inadequate flow of private spending, and enough social support to restore public morale and household buying power. The trouble was that such measures were "impossible" to take because they would have been regarded as tantamount to socialism. Capitalism might have been saved but only by surrendering to the enemy. Thus timid measures were applied—a trickle of government spending and a wholly inadequate cushion of social support—and the depression hung on, arrested but not cured, until World War II arrived and swept away all inhibitions.

Precisely the same paralysis afflicts capitalism today, and I believe that precisely the same resolution awaits it. We would know how to end our present crisis of inflation if war were to be declared: wage and price controls would be instantly emplaced and made effective by high enough taxes on households and businesses to prevent an unsustainable pressure of purchasing power from building up against these dikes. Such controls would undoubtedly bring great aggravations and problems of their own, but they would not be the problem of rampant inflation, just as the remedies of government spending and support introduced new and serious problems, but effectively did away with the problem of massive depression.

As I write these pages, there seems virtually no chance that these "war" measures will be applied in the near-term future. The forces of opinion today are as solidly ranged against such drastic steps as they were against their counterparts in the 1930s, and the rationale in one case is the same as the other: the measures would bring us "socialism." But if inflation continues to accelerate, I expect that capitalism will abandon its halfway measures. It will do so because businessmen and union members alike will come to see that inflation, left to itself, would ultimately injure the system even more than the hated system of controls and higher taxes. (For the very same reasons, I suspect that capitalism in the 1940s would have moved toward full-scale government spending and welfare, if the war had not broken out to legitimate the process.)

Will a capitalism with permanent wage and price controls continue to be capitalism? Of course it will, just as capitalism with a much larger government sector continued to be capitalism. In both cases, to be sure, it is a more "socialistic" capitalism. But that is the contradiction-laden course the system is forced to take to maintain its viability. Capitalism disarms socialism by incorporating some of its elements within itself.

The crucial question, then, is whether inflation will worsen. There are ominous signs that this will be the case—testimony to the self-feeding properties of the inflationary process. To date, only two nations, West Germany and Switzerland, have withstood these cumulative pressures, and these are nations marked by an extraordinary degree of social discipline and cohesion. No other capitalist country has managed to achieve this degree of self-restraint. On the contrary, thirty years of experience with inflation has shown us that the phemonenon has a persistent, although irregular, tendency to quicken.

I think it likely, then, that inflation will drive us in the direction of planning and controls, before we will be forced in that direction by the exigencies of the environment. In that way, the crisis through which capitalism has been passing serves to hasten the advent of the "statism" that seems the direction toward which its internal and external challenges are carrying it. There is a great deal of talk these days about the need to "roll back" the role of government. I suspect this will be a task to be compared with the efforts of King Canute to roll back the sea.

FOUR

The Political Dimension
and "Human Nature"

OUR LENGTHY ANALYSIS OF capitalism
and Western socialism has led to one principal conclusion: the dangers of the human prospect
seem likely to affect the two systems differently in the
short run, but surprisingly alike over a longer time
horizon. As we have seen, this conclusion rests on the
central place which we have assigned to industrial
technology, the source of social and economic pressures that impose common problems on both social
orders, regardless of their different institutions and
ideologies. Beyond that conclusion, however, our
analysis becomes blurred. The logic of socio-economic analysis takes us a certain distance, and then
leaves us with a sense of indeterminacy and incomple-
tion.

The reason is clear enough. Our inquiry has been
entirely conducted by tracing out the "logical dynamics" of a system of profit-seeking firms and individuals, or of efficiency-minded ministries of production.

The Political Dimension and "Human Nature"

What we have omitted has been any consideration of a political dimension—that is, any systematic introduction of the problem of political power, either in terms of the "logical dynamics" of the behavior of nation-states or of those imperatives of behavior or capacities for response that involve the rather ill-defined areas of life we call "political."

The reasons for this omission, in turn, are easy to understand. We live in an age in which the very capacity for socio-economic analysis marks us off from the past. We read with amusement or shock the historical prognoses of the classical historians or political philosophers, into which socio-economic dynamics do not enter at all (for the very good reason that the relevant social systems had not yet evolved) and in which, instead, we find purely political predictions, usually of dynastic rise and fall, and so forth. But however more "scientific" our socio-economic method may seem by comparison, its omission of a political dimension is nonetheless crippling, even fatal, for a comprehension of the human prospect.

For the exercise of political power lies squarely in the center of the determination of that prospect. The resolution of the crises thrust upon us by the social and natural environment can only be found through political action. The dependence of the underdeveloped nations on strong governments has been sufficiently emphasized not to need repetition here. But the very same considerations apply to the nations of the developed world. Here too the most active use

120

of political power will be inescapable, in part as a necessary response to any threats directed at them by the underdeveloped world, in part as the only means to meet and control the challenges of a threatening environment. Certainly the expansive thrust of a market system can be contained and coordinated only by the direct assertion of a greatly expanded domestic national power, as we have indicated; and it is hardly necessary to rehearse the similar conclusions that we reach for industrial socialist nations. As David Calleo and Benjamin Rowland write: "The nation-state may all too seldom speak the voice of reason. But it remains the only serious alternative to chaos."[1]

It is one thing, however, to determine that a political dimension must be added to socio-economic analysis; it is another to provide that dimension. For what is there to be said about the exercise of national power that can compare with the "logical dynamics" of socio-economic reasoning? The classical historians unblushingly likened the course of national history to the life of man, writing of the youth, middle age, and dotage of nations, or took for granted the "human nature" that made the behavior of princely· states as predictable as that of man. But we cannot accept the metaphorical comparisons or the psychological assumptions of these philosophers. What is then left to put in their place? What can be said predictively, or even analytically, about the use of political power?

1. *America and the World Political Economy* (Indiana University Press, 1973), p. 191.

The Political Dimension and "Human Nature"

At the outset we must recognize that there is an aspect of the political dimension that totally eludes our grasp; alas, a vitally important aspect. When we look to the political future to foresee the specific deployment of political power, we are in even greater ignorance than the classicists, who at least thought they knew how men behaved, schemed, and responded with respect to power. We know only that we cannot predict the idiosyncratic behavior of national leaders and therefore cannot foresee the national behavior that is still so much the lengthened shadow of individual leaders. We cannot even predict mass phenomena, such as the "flash points" at which political discontent turns into revolution, or the probabilities that any given regime will muster the support of the people. Thus, over large and critical areas of political behavior, both among and within nations, we are thrown back on our intuitions, hunches, or "wisdom," sometimes presciently, more often not.

But that is not quite an end to it. If the boldest and most far-reaching exercise of political power will be unavoidable over the future, this does more than introduce a random element about which nothing can be said. It also raises the question of whether this exercise of power will be successful, in the sense that it will be accepted by those over whom that power will have to be exercised. *One cannot have political power without political obedience; one cannot have strong government without a sense of national identification.*

How do we know that the use of power, which emerges as such a central necessity for the survival of mankind, will be in fact accepted? What can we say about those traits of political obedience and national identification that we suddenly discover to be the preconditions for the effective mobilization and use of power, whether for evil ends or for life-saving ones?

Here, fortunately, we are not quite in the dark. For the behavioral traits that "permit" the use of political power lie within our scrutiny, even to a certain extent within our predictive capabilities. Therein lies, therefore, the direction in which we must go if we are to introduce the missing political dimension into our inquiry.

Such an effort takes us in the direction of that shadowy concept we call "human nature," but along a very different route from that of the classical historians. We are interested in an examination of man that may throw light on certain attributes of his political behavior. Hence we must begin by focusing our attention on a central fact of human existence—the extended period of helplessness and development through which all human beings must pass and in which the elements of their adult personalities are first molded.[2]

The essential features of this crucial period are

2. For a similarly oriented study, see Harold D. Lasswell, "The Triple-Appeal Principle: A Contribution of Psychoanalysis to Political and Social Science," *American Journal of Sociology*, Jan. 1932.

familiar from the work of Freud and his successors, and can be rapidly summed up. As an infant, still unable to move, the human being experiences (as best we can imagine its scarcely formed consciousness) a sense of infantile omnipotence, in which it "believes" that the world is only an extension of itself, responding to its cries with food, warmth, tactile support, and so on. Moreover, if this "belief" were not in fact based on reality, the infant would perish. Later, as the infant begins to recognize the independent existence of an outer world, it gains the frightening awareness that far from being omnipotent, it is virtually powerless, literally dependent for life itself on the ministration of adults over whom it has no control whatsoever. Later still, as the child seeks to control and direct its physical and psychic energies, it learns to model its behavior on that of adults whose presence is still indispensable and whose wills are irresistible.

In this universal crucible of experience, as we well know, are forged those tendencies in the human personality that later reveal themselves in various sexual, intellectual, aesthetic, moral, and other attitudes. What interests us here, however, are those aspects of the conditioning process that find their vent in the traits of obedience and the capacity for identification—the necessary preconditions for the successful functioning of political institutions in mobilizing individuals for tasks of both peace and war.

The first of these "political" aspects of "human

nature"—the trait of obedience—is surely simple enough to locate in the first few years of experience. What is perhaps less obvious is its expression in adult behavior. The phenomenon to which I wish to call attention is the normal willing acquiescence of men in the exercise of political authority itself. The nature of the "legitimacy" of this authority has been, of course, the object of an extensive discussion, emphasizing such purposes as the preservation of property, the conduct of war, the establishment of law, or, in our own case, the safeguarding of a society threatened by the environment. I have no intention of entering further into this area of "functional" political analysis. Rather, I wish to stress an aspect of political authority that may be obscured by an exclusive concentration on its objective purposes. This latent function is to provide a sense of psychological security by re-creating the accustomed relationships of sub- and superordination to which our long period of helpless dependency has accustomed us.

Certainly we find evidence of this in the ascription of majesty to kings and queens, who are obvious substitutes for our parents, or in the childlike attitudes of mingled resentment and admiration with which the lower orders of society characteristically regard the higher orders, or in the "cult of personality" to which the peoples of the world show such willingness to succumb. Anyone who has seen the wild excitement of a crowd caught up in the adulation of a political

125

leader cannot fail to recognize the rekindling of childhood feelings of awe and obedience in the behavior of these cheering adults.

I am aware, of course, that I tread here on dangerous ground. The experience of childhood is also the source of those drives for self-assertion that contend with obedience, both during and after childhood. Further, it is apparent that the conditioning experience imparts only a very general "tendency" toward obedience—one that finds manifold expressions in adult political behavior, as the most cursory examination of political life reveals in, say, England compared with Italy.

Nor does a stress on the biopsychological underpinnings of political submissiveness deny the importance of other elements which are inextricable from the acquiescence in power. One of these is the presence of force, overtly or covertly employed by the ruling elements to establish and maintain their authority. Another is the differential social conditioning to which different classes in society are exposed. Still another may be the unequal distribution of personality characteristics that lead to power and submission. At still a different level are the hierarchical orderings we observe in many other species.

Nevertheless, a ready admission as to these, or still other, more "positive" reasons for the acceptance of political authority does not explain the phenomenon to which my speculations are addressed. This is the perplexing readiness, even eagerness, with which authority is accepted by the vast majority. An acquiescence, in, or search for, a hierarchical ordering includes not only the lower and middle reaches but also the upper levels of society, who regularly look for "leadership" to someone still higher in the world. Indeed, it finds striking expression in the habit of rulers, including the most dictatorial and absolute, to declare their own "submission" to a will higher than their own, whether it be that of God, of "the people," of some sacred text or doctrine, or of voices audible to themselves alone.

This line of thought has several consequences for the political dimension of the human prospect. To begin with, it offers a substantive basis for our view that the problem of political power exists, not as a mere epiphenomenon of socio-economic relationships, but as a "reality" in its own right whose roots and characteristics can be, at least to some extent, analyzed and applied to the general prognosis for mankind.

In turn this argument has special relevance for several matters we have encountered. One of these is the political outlook for revolutionary socialism, among whose aims is a desire to destratify society to an unprecedented degree. As the example of China illustrates, there is no reason to doubt that impressive changes can be achieved in lessening the social or economic gradations among classes or individuals. But it is useful to consider that the Chinese effort to minimize social and economic hierarchies has taken place within a political framework whose over-all hierarchical structure is as pronounced as that of any society in history. The virtual deification of Mao made China very nearly a personal theocracy, and its striking egalitarian achievements must therefore be viewed in the context of a political order that satisfied the hunger for authority by concentrating it on one remarkable order-bestowing figure. If our speculations are justified, it would follow that revolutionary regimes will be able to perpetuate extreme egalitarian structures only through a succession of leaders endowed with tremendous authority, or else must move in the direction of reestablishing the legitimacy of relations of authority that are now regarded as violations of the revolutionary spirit.

Further, our analysis affords some understanding of the difficulties of democratic governments in managing social tensions. As the histories of the United States, or Switzerland, or modern Scandinavia all illustrate, democracies can provide stable and strong government that assuredly offers some satisfaction for the "political hunger" of mankind. Yet, even in these cases, strong leaders provide a sense of psychological well-being that weak ones do not, so that in moments of crisis and strain demands arise for the exercise of strong-arm rule. As the histories of ancient and modern democracies illustrate, the pressure of political movement in times of war, civil commotion, or general

128

anxiety pushes *in the direction of authority,* not away from it. These tendencies may be short-lived, or may give rise to totalitarian governments that in time collapse, but I do not think that one can deny that these pressures are a persistent fact of political life. One reason for them may indeed lie in the belief— itself perhaps a consequence of the phenomenon of conditioned obedience—that centralized authority will cope with crisis and unrest more "successfully" than less authoritarian structures. But another reason, I venture to suggest, lies in the capacity of powerful "parental" figures, successful or not, to re-create the emotional and psychological custody of one's early years.

I am acutely conscious that this general line of arguments smacks of the worst kind of reactionary ideology: one of the most familiar excuses for dicta- torship is that the masses are "children" and must be treated as such. Nonetheless it would be foolish, as well as hypocritical, not to admit that tendencies toward authoritarian rule seem to be a chronic feature of political life: how many egalitarian revolutions have not ended in the creation of a political establishment every bit as authoritarian as that which they originally displaced? It behooves us therefore to understand this "logic" of political behavior as well as possible, particularly in view of the extraordinary difficulties with which democratic governments will be faced in the coming decades and generations.

Finally, and with great reluctance, I must advance one last implication of my argument. It is customary to recognize, but to deplore, the authoritarian tendencies within civil society, especially on the part of those who, like myself, are the beneficiaries of the freedoms of minimally authority-ridden rule. Yet, candor compels me to suggest that the passage through the gantlet ahead may be possible only under governments capable of rallying obedience far more effectively than would be possible in a democratic setting. If the issue for mankind is survival, such governments may be unavoidable, even necessary. What our speculative analysis provides is not an apologia for these governments, but a basis for understanding the critical support that they may be able to provide for a people who will need, over and above a solution of their difficulties, a mitigation of their existential anxieties.

Let me now advance a second suggestion with regard to the psychological underpinnings of political life. As we have already said, this element concerns the capacity for identification—and in particular *national* identification—which is, like the adult sublimation of childhood obedience, an indispensable precondition for the exercise of political action.

This second political element in "human nature" also finds its origins in the universal conditioning period when the very young child draws its strength and security from those familial figures with whom it mingles its own identity. From this identificatory capacity of the child there flowers, in adult life, an extraordinary array of behavior traits, ranging from the merging of one's self with one's possessions to the capacity for love and sympathy and fellow-feeling. Indeed, the generalized capability of identification is the soil in which are rooted all possibilities of morality.

But we are interested here in the specifically "political" behavior traits that can be traced to this elemental human attribute, and now we find a striking fact. Although the capacity to empathize widens and becomes ever more disciminatingly applied as the child grows older, within every culture of which I have knowledge there seems to be a limit beyond which this general identificatory impulse is blocked. This limit divides those within a society from those beyond it, and demarcates the members of a group among whom a shared concern exists, even though the members may be unknown to one another, from those for whom no such concern is felt.

Once again, it seems possible to trace this otherwise inexplicable fact to the persistence of early childhood attitudes. The child divides the world into two—one comprised of its original family and its subsequent extension of that family; the other of non-familial beings who may exist as human objects but not as human beings with whom an identificatory bond is possible. These same attitudes persist in the political phenomenon of "peoplehood," a phenomenon we find in every culture, ancient and modern. For reasons that we do not fully understand but must accept as a patent fact, nation-states—often with the most hetero-geneous populations—can serve as psychologically valid surrogates for the family and therefore as the beneficiaries of a powerful uniting bond that enables national authorities to concert the actions of diverse individuals. Equally important, nations (or other groups such as tribes or clans) also evidence the limita-tions to the bond of identification, and look upon mem-bers of other states or groups with the same unseeing eye that the child fastens on someone who is merely an object and not a person.

The implications of these remarks for the problem of political prognosis seem clear enough. The feeling of national identity adds another independent under-pinning to the suggestion that the nation-state must be considered as the embodiment of purely political, as well as socio-economic, behavioral forces. Once again this suggestion bears with special relevance on the prospects for revolutionary socialism. For all their socio-economic doctrinal orientation, revolutionary movements most effectively attain their capacity to unite and motivate people when they are welded to the unifying political capabilities of the state. This welding helps us understand the tendency of revolutionary movements, such as the Cuban or Chinese, to infuse their socio-economic teachings with patriotic flavor, together with authoritarian elements of catechism and unimpeachable moral prerogatives. Much of the success of such revolutionary efforts therefore depends on appeals to "primitive" elements—a comment in no way intended to downgrade the actual improvements that these revolutions may bring but to help us understand the nature of the motives on which they are forced to rely.

On a larger scale, the power of the political fantasy in drawing boundaries between those who matter and those who do not carries its disquieting freight for the human prospect in general. For this manifestation of the political element in "human nature" makes it utopian to hope that we will face the global challenges of the future as an international brotherhood of men. If it were possible to imagine the future in terms of the expectations of the 1950s—a "manageable" world in which expert administration would gradually replace the clumsy ignorance of the past—one could hope that the demarcative power of national identification would gradually recede before a kind of international fraternity of administrators and technicians.

The mounting tensions and eventual major transformations that await industrial societies greatly weaken that fond hope. Given the magnitude of the changes that we have sketched out and the competitive struggle for existence that portends, it is unlikely in the extreme that mankind will enjoy a setting in which the identificatory potential within "human nature" can be extended to embrace men and women of other "peoples," or that considerations of a pan-humanistic kind will displace the narrowly familistic basis on which identification is today founded.

For all these forebodings, it is important to recognize that nationalism, despite its potentially vicious application, is not solely a destructive force, and that political identification, with all its problems, is by no means only a dangerous element in "human nature."

Certainly in the underdeveloped world the bond of peoplehood provides an indispensable agency for the mobilization of energies needed to break decisively with the past and to muster the sacrifices needed for the future. And in the developed world, as well, related considerations apply. For when we turn to our own plight, we also face a need to identify with a special group—not one outside our borders, but beyond our reach in time—namely, the generations of the future. A crucial problem for the world of the future will be a concern for generations to come. Where will such a concern arise? Economists speak of the phenomenon of "time discount" as describing the inverted telescope through which humanity looks to the future, estimating the present worth of objects to be enjoyed in the future far below their worth if they could be instantly transferred to the present. This consequent devaluation of the future is generally considered to be an entirely *rational* response to the uncertainties of life. But if we apply this same calculus

134

of "reason" to the human prospect, we face the horrendous possibility that humanity may react to the approach of environmental danger by indulging in a vast fling while it is still possible—a fling entirely justified by the estimation of present enjoyments over future ones. On what private, "rational" considerations, after all, should we make sacrifices now to ease the lot of generations whom we will never live to see?

There is only one possible answer to this question. It lies in our capacity to form a collective bond of identity with those future generations.

Contemporary industrial man, his appetite for the present whetted by the values of a high-consumption society and his attitude toward the future influenced by the prevailing canons of self-concern, has but a limited motivation to form such bonds. There are many who would sacrifice much for their children; fewer who would do so for their grandchildren. *Indeed, it is the absence of just such a bond with the future that casts doubt on the ability of nation-states or socio-economic orders to take now the measures needed to mitigate the problems of the future.*

Is it possible that in another kind of society—one in which it is no longer permissible to indulge in high consumption, perhaps no longer in vogue to set such store by the calculus of selfishness parading as reason —such an identificatory sense could be strengthened? We do not know. Nor do we know to what degree the freedoms and delights of individual self-expression

135

could survive the pressures that would intensify upon the individual in such a community Yet, if the stakes are not those of pleasure but of survival, if the absolute top priority becomes the matter of self-preservation rather than the preservation of the more agreeable aspects of our self-indulgent culture, then I am inclined to believe that the saving element in "human nature" is likely to be that very capability for identification which, in its present political manifestations, also poses some of the most dangerous challenges for the immediate future.

I am quite certain that we have not begun to exhaust the generalizations that can be risked with regard to the political forces at work in history, and I must stress as strongly as possible that I do not have in mind the formulation of an all-embracing "theory" of political behavior. I have entirely omitted for example, the crucial problem of aggression, individual or national, first examined by Freud and since elaborated by many others. I have left unexplored the work of Max Weber or Michels and their followers on the political dynamics of bureaucracy. I have done so in part because the two attributes of "human nature" that I have singled out seem to me to have been neglected, and still more because these attributes seem especially relevant, in a positive sense, to the long-term prospect for survival.

Admittedly, the capacities for submission to power and for identification lack the sense of a clear-cut "dynamics" that is the special characteristic of socio-economic behavior. Yet in calling our attention to the presence of primal elements in the shaping forces of the political future they serve the useful purpose of tempering our expectations with regard to the capacity of socio-economic orders, as such, to cope with the future. That capacity must reckon with the need for—perhaps the ultimate reliance on—welcomed hierarchies of power and strongly felt bonds of peoplehood, to the discomfiture of those who would hope that the challenges of the human prospect would finally banish the thralldoms of authority and ideology and foster the "liberation" of the individual. Our analysis provides a warning that these hopes are not likely to be realized, and that the tensions immanent in socio-economic trends must be worked out within and through the political elements in "human nature." Thus our analysis gives substance to some of the "conservative" reservations with respect to historical change that we find in classical political philosophy, and thereby constitutes a sobering counterbalance to the "radical" expectations that are founded to a large extent on the dynamics of socio-economic change.

The point is important enough to warrant some further elaboration. An essential difficulty in our estimate of the human prospect is the apparent conflict between our intuitive sense of the fixity of "human nature" and our knowledge that behavior can be altered. According to one of the radical tenets, "man makes himself," and is therefore capable of far-reaching changes in his "nature." The conservative takes a more pessimistic view, stressing the presence of a core of "human nature" that offers limits to the possibilities for change. I have sought to avoid the rather vague, and often theological, foreboding that equates this core with "evil," and to suggest that it is better regarded as the psychological substratum of the human personality whose presence we have come to recognize in many areas of behavior and should therefore acknowledge in the sphere of political attitudes as well.

From another view, moreover, I am not so sure that the conservative view is tantamount to a pessimistic view. "Pessimism about man serves to maintain the status quo," writes Leon Eisenberg.[3] Our speculations enlighten us with regard to certain aspects of the "status quo" in *all* societies, such as the susceptibility of men to the submissive requirements of political power and to the fantasies of national "identity" or "purpose," but they do not in themselves offer justification for any particular institutions such as private property, nor do they serve as rationales for the immoral use of political power.

A conservative view of the political element in society must not, therefore, be interpreted as attempt-

3. Leon Eisenberg, "The *Human* Nature of Human Nature," *Science*, April 14, 1972, p. 124.

ing to fix humanity in a vise. Any claim that the quality
of social existence is inexorably determined by the
"nature" of man is refuted out of hand by the most
cursory examination of the range of morality and human sensibility to be found in the various nations of
the world. There remains, nonetheless, the contention
that this plasticity of culture must accommodate itself
in some manner or other to the needs that spring from
man's infant and childhood conditioning, and this does
not permit us to assume that the political structure of
society can accommodate itself to whatever image we
may have of what man should be.

This last consideration is of the essence. The assumption that man ultimately "makes himself" in a
benign manner implies that within the raw stuff of the
human infant there exists some gyroscopic tendency
that will finally guide him, as an adult, in a direction
that will accord with the radical's high moral estimate
of mankind. Otherwise, why should we not conclude
that the self-made man, stripped of all his false
consciousness, divested of the delusions and fantasies
that have misled him, will settle into a state of utter
existential despair, or relapse into a suicidal solipsism?
Indeed, why not conclude that before the terrifying
truth of mortal finitude each man must shed the frail
moral teachings of the past and finish his life in an orgy
of self-indulgence that knows no bounds? As we have
suggested, that truly pessimistic possibility must look
for its refutation to the persistent promptings of a

139

portion of man's being that he does not "make," but that makes him. In this regard it is worth reflecting that the hideous visions of man's future in Huxley's *Brave New World,* or Orwell's *1984* are both based on the premise of the unlimited plasticity and malleability of the human species.

It is possible, of course, that in the future men may be so altered in their genetic characters, or nurtured in such carefully planned circumstances, that the "class" or "patriotic" attributes of political life would disappear because they no longer answered to an inner need. But at this juncture in history, our attention had better be focused on what men are likely to be, rather than on what they could eventually become. The human prospect forces us to deal with human change within an indeterminate, but not indefinite, time period, and speculations as to the degree of potential change must give way before the degree of change that is imaginable within that period.

So far as the genetic question is concerned, the time required for change is very long indeed, unless we discover chemical means of altering human behavior and apply these on a global scale—a prospect still happily well beyond reach. Writing in a symposium on behavior, E. O. Wilson presents the following "optimistic" estimate:

[T]here is every justification from both genetic theory and experiments on animal species to suppose that rapid behavioral evolution is at least a possibility in man. By rapid

140

I mean significant alteration in, say, emotional and intellectual traits within no more than ten generations—or about 300 years.[4]

Unfortunately, three hundred years, however rapid in the eye of the anthropologist, is hopelessly slow for the challenges now gathering on the horizon. Moreover, Wilson does not specify the changes in social institutions that might be required to bring about the accelerated evolution in the direction of social improvement.

As for the rapidity with which these institutional changes can work their effect on behavior, we face the problem of the natural "inertia" of the human condition, an inertia ascribable not only to the presence of a stubbornly persisting substratum of psychological needs but also to the laggard pace of change in the family setting through which those needs are gradually shaped into the attitudes of adult behavior. In his sympathetic but critical summary of Marx's view of man, Bertell Ollman vividly describes this process:

People acquire most of their personal and class characteristics in childhood. It is the conditions operating then, transmitted primarily by the family, which makes them what they are, at least as regards basic responses; and, in most cases, what they are will vary very little over their lives. Thus, even where the conditions people have been brought up in change by the time they reach maturity, their

4. E. O. Wilson, "Competitive and Aggressive Behavior," in *Man and Beast* (Smithsonian Institution, 1971), p. 207.

characters still reflect the situation which has passed on. If Marx had studied the family more closely, surely he would have noticed that as a factory for producing character it is invariably a generation or more behind the times, producing people who, tomorrow, will be able to deal with yesterday's problems.[5]

I do not raise these considerations to dismiss the possibility of dramatic transformations in social organization, such as we have seen in China. Indeed, I am persuaded that changes of at least this degree of penetration and revolutionary impact will be required within the time span with which our examination has been concerned. My analysis leads me, rather, to reiterate that these behavioral alterations, much as those in China, will have to allow for, or build on, recalcitrant elements in the human personality, including the two that I have singled out for emphasis, namely, the "hunger" for political authority and the "fantasy" of political identification. Further, it is not genetic evolution or cumulative amelioration in rearing that is likely to be the crucial implementing factor in affecting the behavioral reorientations of the "post-industrial" future, but the use of those primal elements on which political power rests—a belief for which, once again, the Chinese experience provides supporting evidence.

I am all too aware that these conclusions may bring dismay to many whom I consider my friends and comfort to many whom I consider my foes. To suggest

5. Bertell Ollman, *Alienation: Marx's Conception of Man in Capitalist Society* (Cambridge University Press, 1971), p. 241.

that political power and hierarchy serve a supportive function in society plays directly into the hands of those who applaud the "orderliness" of authoritarian or dictatorial governments. To find a reason for the appeal of nonrational political beliefs is to encourage those who advocate irresponsible political programs. To stress the psychological roots of peoplehood is to weaken the cause of whose who seek to overcome the curse of racism and xenophobia.

If I nonetheless publish these thoughts, with all their potential mischievousness, it is for two reasons. The first is that the weakest part of the humanitarian outlook, both philosophically and pragmatically, has been its inability or unwillingness to come to grips with certain obdurate human characteristics. As a result we find buried within "humanist" appeals a conception of human nature that is often as reactionary, in the sense of ascribing an inherent element of evil to man, as that of the most unthinking conservative. Let me cite this example from a contemporary radical publication:

> In the most profound sense, the proletariat has not one enemy but two—the ruling class and itself. In the absence of a humanizing militancy and a militant humanism, in the absence of a fierce common hatred for the common enemy, and a fiercer common love for the proletariat as a whole, history will degenerate into barbarism.[6]

Extended commentary hardly seems necessary. The encouragement of aggressive impulses (militancy,

6. "The Making of Socialist Consciousness," by the editors of *Socialist Revolution* (1970), reprinted in *The Capitalist System*, eds. R.C. Edwards, Reich, and Weiskopf (Prentice-Hall, 1972), p. 505.

fierce hatred, fiercer love), the dehumanization implicit in the admonition to "love" the proletariat "as a whole", and above all the view of man as engaged in a struggle to the death with himself, open this view to a critique as scathing as any that could be directed against a "bourgeois" conception of humanity. If radicalism is to go to the roots, as the term implies, it must be prepared to examine the "nature" of man in ways much more courageous and much less pietistic than those it uses in the name of "humanism." Only on such a basis can it hope to build ideas and programs that may be able to withstand the tempest of events whose source lies, both as challenge and response, within men themselves.

My second reason for advancing these views relates to the first. I have tried to take the measure of man as a creature of his socio-economic arrangements and his political bonds. It may be that from some other perspective the prospect for collective human adaptation would seem brighter. But from the vantage point of this book, a failure to recognize the limitations and difficulties of our capacities for response would only build an architecture of hope on false beliefs.

AFTERWORD

THE PROBLEM OF POWER and its relation to "human nature"—that elusive but unavoidable idea—occupies the center of this chapter.

There is nothing I can add (and nothing I wish to subtract) from my speculations as to its psychic roots. There is, however, an aspect of the problem that I would like to raise in this Afterword, calling attention to the sharply opposing ways in which power is viewed from the perspectives of the Right and the Left.

The Right has always recognized that power gives shape and force to historical events, but it has typically regarded the exercise of power as nothing but the expression of an invariant human propensity. Therefore conservative writers emphasize the sheer *fact* of domination as the basic, repeating pattern of history. "Plus ça change, plus c'est la même chose," was La Rochefoucault's cynical way of putting it; "All power tends to corrupt, and absolute power tends to corrupt absolutely" was Lord Acton's more genteel statement of the same thought; and in still more abstract fashion Plato and Aristotle, Machiavelli and Hobbes advanced the same generalization: domination is an inescapable tendency in history because man is a dominating animal by his nature (and I use the masculine pronoun advisedly).[1]

I need hardly add that such a view has a prima facie cogency, for domination in one form or another can indeed be traced through the long human narra-

---

1. Recent work by sociobiologist E. O. Wilson (*On Human Nature*, Harvard Univ. Press, 1978) has lent support to the genetic basis of social orderings. Wilson's work has been attacked as ideological rather than scientific; I suspect the issue will be debated for a long time. For interesting reviews see David Pilbeam, "Toward a Concept of Man," *Natural History*, Feb. 1979, pp. 100f, and N.J. Mackintosh, "A Proffering of Underpinnings," *Science*, May 18, 1979, pp. 735–37.

tive. Moreover, because again and again efforts to overthrow domination in one form end up welcoming, or at least accepting, domination in another form, the conservative view suggests that it is founded on something more substantial than just a jaundiced view of humankind. Indeed, by preparing us for the worst, the conservative perspective helps us to forestall it, by insisting on constitutional barriers, legal barriers, and the like—frail defenses, perhaps, against the assertion of unrestrained power, but the only defenses we possess.

What the conservative view fails to recognize, however, is that power is rarely a wild card in the game of politics, but a trump suit held by the ruling interests of society. Power has a systematic aspect to it that the conservative ignores—an aspect that brings material, and social rewards to certain elements of society while withholding them from others. Thus the identification of domination as the expression of an eternal aspect of human nature, however useful in alerting us to such a propensity and in preparing us to prevent its abuses, nonetheless usually serves as an apology for the particular structure of power in a given society by diverting attention from its beneficiaries and its victims.

The view of the Left is exactly the opposite. The Left also sees power as a continuing theme of history, but unlike the Right, it sees power as always deployed in the favor of privileged groups against unprivileged ones. When Marx and Engels declare in the *Manifesto* that the history of all previously existing society is the history of class struggles, they are placing this systematic use of power at the forefront of historical analysis, and directing our attention at the very question of *class* domination from which the conversatives avert their eyes.

Yet there are also striking weaknesses within the Left's view of power. One weakness is that the Left fastens on exploitation by ruling socio-economic classes, and ignores the use of power by men to dominate women, or by light-skinned peoples to dominate dark-skinned ones; or by intellectual, religious, or political elites to dominate the masses for reasons that are not primarily material. Moreover, even within its paradigm of domination for the sake of socio-economic privilege, it does not ask what actual pleasures or purposes are served by the exercise of domination or exploitation.

Why do men seek wealth or status? It is only by asking this "simple" question that the ultimate gains and appeals of power can be uncovered and understood. Moreover, it is only by tracing these gains and appeals that the Left can explain not alone the recurrent abuse of power in history, but also the attitudes of those over whom power is exercised. Thus, both the consequences and the possibilities of revolution will remain impenetrable mysteries until the act of, and the acquiescence in, domination have been traced to their sources within the psyche—that is, within human nature, as it is formed and shaped in all societies.

Such an effort is indispensable if the Left is to escape from its besetting weakness—its failure to anticipate the political disasters that have been the curse of Marxist socialism in our time. As long as the problem of power is left unexplored and even unacknowledged, at best shrugged aside as a manifestation of class societies that will disappear under the dispensation of an undefined "participatory democracy," political catastrophe will dog the heels of all socialist movements.

To state this is in no way to belittle ambitious programs or lofty ideals for human betterment. It is only to warn that human nature must be given its respectful due if these programs and ideals are not to end in terrible surprises and unforeseen miscarriages. The hope is that the psychological insights of conservative thought can be welded to the penetrative social analysis of radical investigation. Here a few promising starts have been made, but there is a vast deal to be achieved.[2]

2. Two path-breaking studies are Dorothy Dinnerstein, *The Mermaid and the Minotaur* (Harper, 1977), an exploration of the problem of male-female domination; and Joel Kovel, *White Racism* (Random House, 1971), a Marxian and Freudian exploration of the roots of racism.

FIVE

Final Reflections on the Human Prospect

WHAT IS NEEDED NOW is a summing up of the human prospect, some last reflections on its implications for the present and future alike.

The external challenges can be succinctly reviewed. We are entering a period in which rapid population growth, the presence of obliterative weapons, and dwindling resources will bring international tensions to dangerous levels for an extended period. Indeed, there seems no reason for these levels of danger to subside unless population equilibrium is achieved and some rough measure of equity reached in the distribution of wealth among nations, either by great increases in the output of the underdeveloped world or by a massive redistribution of wealth from the richer to the poorer lands.

Whether such an equitable arrangement can be reached—at least within the next several generations—is open to serious doubt. Transfers of adequate magnitude imply a willingness to redistribute income

149

internationally on a more generous scale than the advanced nations have evidenced within their own domains. The required increases in output in the backward regions would necessitate gargantuan applications of energy merely to extract the needed resources. It is uncertain whether the requisite energy-producing technology exists, and, more serious, possible that its application would bring us to the threshold of an irreversible change in climate as a consequence of the enormous addition of man-made heat to the atmosphere.

It is this last problem that poses the most demanding and difficult of the challenges. The existing pace of industrial growth, with no allowance for increased industrialization to repair global poverty, holds out the risk of entering the danger zone of climatic change in as little as three or four generations. If that trajectory is in fact pursued, industrial growth will then have to come to an immediate halt, for another generation or two along that path would literally consume human, perhaps all, life. That terrifying outcome can be postponed only to the extent that the wastage of heat can be reduced, or that technologies that do not add to the atmospheric heat burden—for example, the use of solar energy—can be utilized. The outlook can also be mitigated by redirecting output away from heat-creating material outputs into the production of "services" that add only trivially to heat.

All these considerations make the designation of a timetable for industrial deceleration difficult to construct. Yet, under any and all assumptions, one irrefutable conclusion remains. The industrial growth process, so central to the economic and social life of capitalism and Western socialism alike, will be forced to slow down, in all likelihood within a generation or two, and will probably have to give way to decline thereafter. To repeat the words of the text, "whether we are unable to sustain growth or unable to tolerate it," the long era of industrial expansion is now entering its final stages, and we must anticipate the commencement of a new era of stationary total output and (if population growth continues or an equitable sharing among nations has not yet been attained) declining material output per head in the advanced nations.

These challenges also point to a certain time frame within which different aspects of the human prospect will assume different levels of importance. In the short run, by which we may speak of the decade immediately ahead, no doubt the most pressing questions will be those of the use and abuse of national power, the vicissitudes of the narrative of political history, perhaps the short-run vagaries of the economic process, about which we have virtually no predictive capability whatsoever. From our vantage point today, another crisis in the Middle East, further Vietnams or Czechoslovakias, inflation, severe economic malfunction—or their avoidance—are sure to exercise the primary influence over the quality of existence, or even over the possibilities for existence.

In a somewhat longer time frame—extending perhaps for a period of a half century—the main shaping force of the future takes on a different aspect. Assuming that the day-to-day, year-to-year crises are surmounted in relative safety, the issue of the relative resilience and adaptive capabilities of the two great socio-economic systems comes to the fore as the decisive question. Here the properties of industrial socialism and capitalism as ideal types seem likely to provide the parameters within which and by which the prospect for man will be formed. We have already indicated what general tendencies seem characteristic of each of these systems, and the advantages that may accrue to socialist—that is, planned and probably authoritarian social orders—during this era of adjustment.

In the long run, stretching a century or more ahead, still a different facet of the human prospect appears critical. This is the transformational problem, centered in the reconstruction of the material basis of civilization itself. In this period, as indefinite in its boundaries but as unmistakable in its mighty dimensions as a vast storm visible on the horizon, the challenge devolves upon those deep-lying capabilities for political change whose roots in "human nature" have been the subject of our last chapter.

It is the challenges of the middle and the long run that command our attention when we speculate about the human prospect, if only because those of the short run defy our prognostic grasp entirely. It seems

152

unnecessary to add more than a word to underline the magnitude of these still distant problems. No developing country has fully confronted the implications of becoming a "modern" nation-state whose industrial development must be severely limited, or considered the strategy for such a state in a world in which the Western nations, capitalist and socialist both, will continue for a long period to enjoy the material advantages of their early start. Within the advanced nations, in turn, the difficulties of adjustment are no less severe. No capitalist nation has as yet imagined the extent of the alterations it must undergo to attain a viable stationary socio-economic structure, and no socialist state has evidenced the needed willingness to subordinate its national interests to supra-national ones.

To these obstacles we must add certain elements of the political propensities in "human nature" that stand in the way of a rational, orderly adaptation of the industrial mode in the directions that will become increasingly urgent as the distant future comes closer. There seems no hope for rapid changes in the human character traits that would have to be modified to bring about a peaceful, organized reorientation of life styles. Men and women, much as they are today, will set the pace and determine the necessary means for the social changes that will eventually have to be made. The drift toward the strong exercise of political power—a movement given its initial momentum by the need to exercise a much wider and deeper administration of both production and consumption—is likely to attain added support from the psychological insecurity that will be sharpened in a period of unrest and uncertainty. The bonds of national identity are certain to exert their powerful force, mobilizing men for the collective efforts needed but inhibiting the international sharing of burdens and wealth. The myopia that confines the present vision of men to the short-term future is not likely to disappear overnight, rendering still more difficult a planned and orderly retrenchment and redivision of output.

Therefore the outlook is for what we may call "convulsive change"—change forced upon us by external events rather than by conscious choice, by catastrophe rather than by calculation. As with Malthus's much derided but all too prescient forecasts, nature will provide the checks, if foresight and "morality" do not. One such check could be the outbreak of wars arising from the explosive tensions of the coming period, which might reduce the growth rates of the surviving nation-states and thereby defer the danger of industrial asphyxiation for a period. Alternatively, nature may rescue us from ourselves by what John Platt has called a "storm of crisis problems."[1] As we breach now this, now that edge of environmental tolerance, local disasters—large-scale fatal urban tempera-

1. John Platt, "What We Must Do," *Science*, Nov. 28, 1969, p. 1115.

ture inversions, massive crop failures, resource shortages—may also slow down economic growth and give a necessary impetus to the piecemeal construction of an ecologically and socially viable social system.

Such negative feedbacks are likely to exercise an all-important dampening effect on a crisis that would otherwise in all probability overwhelm the slender human capabilities for planned adjustment to the future. However brutal these feedbacks, they are apt to prove effective in changing our attitudes as well as our actions, unlike appeals to our collective foresight, such as the exhortations of the Club of Rome's *Limits to Growth*, or the manifesto of a group of British scientists calling for an immediate halt to growth.[2] The problem is that the challenge to survival still lies sufficiently far in the future, and the inertial momentum of the present industrial order is still so great, that no substantial voluntary diminution of growth, much less a planned reorganization of society, is today even remotely imaginable. What leader of an underdeveloped nation, particularly one caught up in the exhilaration of a revolutionary restructuring of society, would call a halt to industrial activity in his impoverished land? What capitalist or socialist nation would put a ceiling on material output, limiting its citizens to the well-being obtainable from its present volume of production?

Thus, however admirable in intent, impassioned

2. "Blueprint for Survival," *The Ecologist*, Jan. 1972.

polemics against growth are exercises in futility today. Worse, they may even point in the wrong direction. Paradoxically, perhaps, the priorities for the present lie in the temporary encouragement of the very process of industrial advance that is ultimately the mortal enemy. In the backward areas, the acute misery that is the potential source of so much international disruption can be remedied only to the extent that rapid improvements are introduced, including that minimal infrastructure needed to support a modern system of health services, education, transportation, fertilizer production, and the like. In the developed nations, what is required at the moment is the encouragement of technical advances that will permit the extraction of new resources to replace depleted reserves of scarce minerals, new sources of energy to stave off the collapse that would occur if present energy reservoirs were exhausted before substitutes were discovered, and, above all, new techniques for the generation of energy that will minimize the associated generation of heat.

Thus there is a short period left during which we can safely continue on the present trajectory. It is possible that during this period a new direction will be struck that will greatly ease the otherwise inescapable adjustments. The underdeveloped nations, making a virtue of necessity, may redefine "development" in ways that minimize the need for the accumulation of capital, stressing instead the education and vitality of their citizens. The possibilities of such an historic step would be much enhanced were the advanced nations to lead the way by a major effort to curtail the enormous wastefulness of industrial production as it is used today. If these changes took place, we might even look forward to a still more desirable redirection of history in a diminution of scale, a reduction in the size of the human community from the dangerous level of immense nation-states toward the "polis" that defined the appropriate reach of political power for the ancient Greeks.

All these are possibilities, but certainly not probabilities. The revitalization of the polis is hardly likely to take place during a period in which an orderly response to social and physical challenges will require an increase of centralized power and the encouragement of national rather than communal attitudes. The voluntary abandonment of the industrial mode of production would require a degree of self-abnegation on the part of its beneficiaries—managers and consumers alike—that would be without parallel in history. The redefinition of development on the part of the poorer nations would require a prodigious effort of will in the face of the envy and fear that Western industrial power and "affluence" will arouse.

Thus in all likelihood we must brace ourselves for the consequences of which we have spoken—the risk of "wars of redistribution" or of "preemptive seizure," the rise of social tensions in the industrialized

157

nations over the division of an ever more slow-growing or even diminishing product, and the prospect of a far more coercive exercise of national power as the means by which we will attempt to bring these disruptive processes under control.

From that period of harsh adjustment, I can see no realistic escape. Rationalize as we will, stretch the figures as favorably as honesty will permit, we cannot reconcile the requirements for a lengthy continuation of the present rate of industrialization of the globe with the capacity of existing resources or the fragile biosphere to permit or to tolerate the effects of that industrialization. Nor is it easy to foresee a willing acquiescence of humankind, individually or through its existing social organizations, in the alterations of lifeways that foresight would dictate. If then, by the question "Is there hope for man?" we ask whether it is possible to meet the challenges of the future without the payment of a fearful price, the answer must be: No, there is no such hope.

At this final stage of our inquiry, with the full spectacle of the human prospect before us, the spirit quails and the will falters. We find ourselves pressed to the very limit of our personal capacities, not alone in summoning up the courage to look squarely at the dimensions of the impending predicament, but in finding words that can offer some plausible relief in a situation so bleak. There is now nowhere to turn other than to those private beliefs and disbeliefs that guide each of us through life, and whose disconcerting presence was the first problem with which we had to deal in appraising the prospect before us. I shall therefore speak my mind without any pretense that the words I am about to write have any basis other than those subjective promptings from which I was forced to begin and in which I must now discover whatever consolation I can offer after the analysis to which they have driven me.

At this late juncture I have no intention of sounding a call for moral awakening or for social action on some unrealistic scale. Yet, I do not intend to condone, much less to urge, an attitude of passive resignation, or a relegation of the human prospect to the realm of things we choose not to think about. Avoidable evil remains, as it always will, an enemy that can be defeated; and the fact that the collective destiny of man portends unavoidable travail is no reason, and cannot be tolerated as an excuse, for doing nothing. This general admonition applies in particular to the intellectual elements of Western nations whose privileged role as sentries for society takes on a special importance in the face of things as we now see them. It is their task not only to prepare their fellow citizens for the sacrifices that will be required of them but to take the lead in seeking to redefine the legitimate boundaries of power and the permissible sanctuaries of freedom, for a future in which the exercise of power

159

must inevitably increase and many present areas of freedom, especially in economic life, be curtailed.

Let me therefore put these last words in a somewhat more "positive" frame, offsetting to some degree the bleakness of our prospect, without violating the facts or spirit of our inquiry. Here I must begin by stressing for one last time an essential fact. The human prospect is not an irrevocable death sentence. It is not an inevitable doomsday toward which we are headed, although the risk of enormous catastrophes exists. The prospect is better viewed as a formidable array of challenges that must be overcome before human survival is assured, before we can move *beyond doomsday*. These challenges can be overcome—by the saving intervention of nature if not by the wisdom and foresight of man. The death sentence is therefore better viewed as a contingent life sentence—one that will permit the continuance of human society, but only on a basis very different from that of the present, and probably only after much suffering during the period of transition.

What sort of society might eventually emerge? As I have said more than once, I believe the long-term solution requires nothing less than the gradual abandonment of the lethal techniques, the uncongenial lifeways, and the dangerous mentality of industrial civilization itself. The dimensions of such a transformation into a "post-industrial" society have already been touched upon, and cannot be greatly elaborated here:

in all probability the extent and ramifications of change are as unforeseeable from our contemporary vantage point as present-day society would have been unimaginable to a speculative observer a thousand years ago.

Yet I think a few elements of the society of the post-industrial era can be discerned. Although we cannot know on what technical foundation it will rest, we can be certain that many of the accompaniments of an industrial order must be absent. To repeat once again what we have already said, the societal view of production and consumption must stress parsimonious, not prodigal, attitudes. Resource-consuming and heat-generating processes must be regarded as necessary evils, not as social triumphs, to be relegated to as small a portion of economic life as possible. This implies a sweeping reorganization of the mode of production in ways that cannot be foretold, but that would seem to imply the end of the giant factory, the huge office, perhaps of the urban complex.

What values and ways of thought would be congenial to such a radical reordering of things we also cannot know, but it is likely that the ethos of "science," so intimately linked with industrial application, would play a much reduced role. In the same way, it seems probable that a true post-industrial society would witness the waning of the work ethic that is also intimately entwined with our industrial society. As one critic has pointed out, even Marx,

161

despite his bitter denunciation of the alienating effects of labor in a capitalist milieu, placed his faith in the presumed "liberating" effects of labor in a socialist society, and did not consider a "terrible secret"—that even the most creative work may be only "a neurotic activity that diverts the mind from the diminution of time and the approach of death."[3]

It is therefore possible that a post-industrial society would also turn in the direction of many pre-industrial societies—toward the exploration of inner states of experience rather than the outer world of fact and material accomplishment. Tradition and ritual, the pillars of life in virtually all societies other than those of an industrial character, would probably once again assert their ancient claims as the guide to and solace for life. The struggle for individual achievement, especially for material ends, is likely to give way to the acceptance of communally organized and ordained roles.

This is by no means an effort to portray a future utopia. On the contrary, many of these possible attributes of a post-industrial society are deeply repugnant to my twentieth-century temper as well as incompatible with my most treasured privileges. The search for scientific knowledge, the delight in intellectual heresy, the freedom to order one's life as one pleases, are not likely to be easily contained within the tradition-

3. John Diggins, "Thoreau, Marx, and the Riddle of Alienation," *Social Research*, Winter 1973, p. 573.

oriented, static society I have depicted. To a very great degree, the public must take precedence over the private—an aim to which it is easy to give lip service in the abstract but difficult for someone used to the pleasures of political, social, and intellectual freedom to accept in fact.

These are all necessarily prophetic speculations, offered more in the spirit of providing some vision of the future, however misty, than as a set of predictions to be "rigorously" examined. In these half-blind gropings there is, however, one element in which we can place credence, although it offers uncertainty as well as hope. This is our knowledge that some human societies have existed for millennia, and that others can probably exist for future millennia, in a continuous rhythm of birth and coming of age and death, without pressing toward those dangerous ecological limits, or engendering those dangerous social tensions, that threaten present-day "advanced" societies. In our discovery of "primitive" cultures, living out their timeless histories, we may have found the single most important object lesson for future man.

What we do not know, but can only hope, is that future man can rediscover the self-renewing vitality of primitive culture without reverting to its levels of ignorance and cruel anxiety. It may be the sad lesson of the future that no civilization is without its pervasive "malaise," each expressing in its own way the ineradicable fears of the only animal that contemplates its own death, but at least the human activities expressing that malaise need not, as is the case in our time, threaten the continuance of life itself.

All this goes, perhaps, beyond speculation to fantasy. But something more substantial than speculation or fantasy is needed to sustain men through the long trials ahead. For the driving energy of modern man has come from his Promethean spirit, his nervous will, his intellectual daring. It is this spirit that has enabled him to work miracles, above all to subjugate nature to his will, and to create societies designed to free man from his animal bondage.

Some of that Promethean spirit may still serve us in good stead in the years of transition. But it is not a spirit that conforms easily with the shape of future society as I have imagined it; worse, within that impatient spirit lurks one final danger for the years during which we must watch the approach of an unwanted future. This is the danger that can be glimpsed in our deep consciousness when we take stock of things as they now are: the wish that the drama run its full tragic course, bringing man, like a Greek hero, to the fearful end that he has, however unwittingly, arranged for himself. For it is not only with dismay that Promethean man regards the future. It is also with a kind of anger. If after so much effort, so little has been accomplished; if before such vast challenges, so little is apt to be done—then let the drama proceed to its finale, let mankind suffer the end it deserves.

Such a view is by no means the expression of only a few perverse minds. On the contrary, it is the application to the future of the prevailing attitudes with which our age regards the present. When men can generally acquiesce in, even relish, the destruction of their living contemporaries, when they can regard with indifference or irritation the fate of those who live in slums, rot in prison, or starve in lands that have meaning only insofar as they are vacation resorts, why should they be expected to take the painful actions needed to prevent the destruction of future generations whose faces they will never live to see? Worse yet, will they not curse these future generations whose claims to life can be honored only by sacrificing present enjoyments; and will they not, if it comes to a choice, condemn them to nonexistence by choosing the present over the future?

The question, then, is how we are to summon up the will to survive—not perhaps in the distant future, where survival will call on those deep sources of imagined human unity, but in the present and near-term future, while we still enjoy and struggle with the heritage of our personal liberties, our atomistic existences.

At this last moment of reflection another figure from Greek mythology comes to mind. It is that of Atlas, bearing with endless perseverance the weight of the heavens in his hands. If mankind is to rescue life, it must first preserve the very will to live, and thereby rescue the future from the angry condemnation of the

165

present. The spirit of conquest and aspiration will not provide the inspiration it needs for this task. It is the example of Atlas, resolutely bearing his burden, that provides the strength we seek. If, within us, the spirit of Atlas falters, there perishes the determination to preserve humanity at all cost and any cost, forever.

But Atlas is, of course, no other but ourselves. Myths have their magic power because they cast on the screen of our imaginations, like the figures of the heavenly constellations, immense projections of our own hopes and capabilities. We do not know with certainty that humanity will survive, but it is a comfort to know that there exist within us the elements of fortitude and will from which the image of Atlas springs.

AFTERWORD

THERE IS a fearful question posed at the conclusion of this chapter: Will mankind have the fortitude to undertake Atlas's task? That is a question I shall postpone to the Postscript that follows. Instead, I would like to begin this last reconsideration by posing a much more down-to-earth problem. It is whether we can gain a better reading of the timetable of events than when I wrote this chapter originally.

I believe we can today divide the fairly near-term

future into two periods, the first marked by its continued emphasis on growth and on "business as usual," the second by its awareness of the dangers of growth and its conscious search for a new framework of socio-economic organization. Moreover, I think we can locate with a fair degree of plausibility where the zone of demarcation lies. Three separate indicators point to a period roughly twenty-five years ahead—about the span we call a generation—that separates the first from the second.

The first of these three indicators is provided by the outlook for petroleum prices. The price of oil today is not set by a free market, but by the action of the OPEC cartel. Although the price is very high by historic standards, it is the prop of supply, rather than the pull of demand, that essentially establishes the level of prices. Nonetheless, demand is steadily increasing as the world uses ever more oil, and at some date in the future oil prices will begin steadily to rise, *not because of OPEC's actions, but because demand will be outrunning supply.* From that time on, oil will become a truly "scarce" resource, not one whose scarcity is the consequence of the monopoly power of its suppliers.

The date of this "cross-over" of demand and supply will depend, of course, on the rates at which we consume oil and at which we discover new reservoirs of it. But what is surprising is how little difference it makes whether we apply optimistic or pessimistic estimates to these determining factors. This is because we continue to increase our consumption of oil at exponential rates, and all exponentially growing processes use up resources at bewildering speeds, as we saw in the Afterword to Chapter Two. According to the present estimates of the exponential growth in the use of oil and the availability of oil reserves, we will have reached cross-over by the mid 1980s under pessimistic assumptions and by the late 1990s under optimistic ones. Thus only fifteen years separates the best case from the worst one.[1] If we arbitrarily add another ten years to the best case, we would put the date of cross-over in the middle of the first decade of the next century, just twenty-five years ahead.

The point, of course, is not to try to fix the exact year in which cross-over will occur. It is rather to use cross-over, with its prospect of ever-rising prices thereafter, to establish a plausible time frame for the period available to the industrial world for a shift away from a petroleum-based technology to a coal, nuclear, and solar-based one. A period of twenty-five years seems like a realistic estimate of the period during which we must make the transition from the present structure of production to another, as yet undeveloped.

A second independent estimate stems from a study of the world's interlocked needs and requirements conducted by Nobelist Wassily Leontief for the

---

1. See Flower, *op. cit.*, and Issawi, *op. cit.*

United Nations. Leontief's study of the world economy was undertaken to inquire whether it was possible
to anticipate another twenty-five years of growth at
rates similar to those of the previous decades. Leontief's answer was a very cautious yes. It was cautious
because it made clear the staggering requirements for
such a continued rate of global expansion—food outputs quadrupled and mineral tonnages quintupled—
and because it recognized that the scale of investments
—fertilizers, irrigation, transportation networks,
energy—to attain these goals would place an unprecedented strain on both rich and poor nations.
Investment requirements for the underdeveloped
world were estimated to range up to 40 percent of
their gross national products, levels that have been
reached only under the most extreme warlike
conditions.[2]

Thus Leontief's study allows us to imagine a
world continuing its present growth path for twenty-
five years, if it can marshal the political and social will
to do so. What that implies by way of a drift toward
authoritarian regimes is itself sobering enough. But the
relevant consideration for our study is something else.
What happens after the twenty-fifth year? How is the
growth trajectory to be maintained, if it will take such
exhausting efforts to sustain it during the next quarter
century?

2. Wassily Leontief, *The Future of the World Economy* (Oxford
University Press, 1977), pp. 4,5,11.

The plain answer, I believe, is that it will *not* be maintained, and that Leontief's study allows us to picture the greatest possible attainment of the world's productive capacity, not the attainment it will most likely achieve. I am prepared to hope that world output will continue to grow for another twenty-five years, but it seems clear that the growth rate at the end of the period will be much lower than at the beginning. These twenty-five years of slowing growth will give us roughly one more doubling of output per capita. Another doubling seems very difficult to imagine.

Third, there is the benchmark provided by warning signals about our intervention into the biosphere, a matter we have looked into in the Afterword to Chapter II. There we saw that we are already invading the life support system of the green-blue film of water and air to the point where we threaten its indispensable life support capacities. And the scale of intervention steadily grows, partly as the result of the sheer accumulation of the mass of industrial effluents; partly because technology itself becomes ever more powerful and potentially hazardous, nuclear energy being, of course, the primary instance.

These warning signals do not in themselves establish any datelines—perhaps we could continue to skate on thin ice indefinitely. But they suggest that the ice is already creaking ominously, and that the time before it cracks cannot be indefinitely postponed. It is no more than a guess that the environment also establishes a twenty-five year warning period, but I do not think it is a groundless, or irresponsible guess.

And so I think a somewhat clearer timetable of change can be projected into the medium-term future. For roughly twenty-five years, the industrialized capitalist and socialist worlds can probably continue along their present growth paths, although energy constraints and environmental dangers will no doubt be enforcing a gradual slowing down of growth rates. Thereafter, I think we can expect not only a much more pronounced braking of growth, but a general recognition that the possibilities for expansion are limited, and that social and economic life must be maintained within fixed, rather than outward-moving, material boundaries.

But the real problem to be faced in this final assessment is not that of establishing a timetable for resource or environmental frictions. It is taking the measure of the institutional and attitudinal changes that will be required of future generations, and of weighing the various means by which society can enforce whatever adaptational or transformational changes will be necessary for survival.

Here I have little to add to the discussions of previous chapters, except to warn that the problems of tomorrow must be solved before we can address ourselves to those of the day after tomorrow. This has special relevance to those whose impulse, when con-

fronted with the severe demands of the human prospect, is to seek immediately to junk our present way of
life and to establish those small-scale, self-sufficient
communities that beckon to us as a radically different,
and far preferable, alternative to present day industrial
civilization.

Perhaps some day these visions can be realized.
But first humanity must be rescued from its exposed
and dangerous plight. This requires action on the
grand scale, not on the small scale. Mankind lives in
immense urban complexes and these must be sus-
tained and provisioned for a long time. Structures of
production blocks long and months deep cannot be
quickly broken down into pocket-sized miniatures,
and can only be abandoned at the risk of social collapse. Dangerous military stores must be guarded.
Hospitals must be maintained. The network of com-
munication cannot be allowed to come apart. All this
will necessitate central authority as the condition for
survival. Pockets of small-scale communities may be
established, but they will be parasitic to, not genuine
alternatives for, the centralized regime that will be
struggling to redesign society.

Given this mighty task, we must think of alternatives to the present order in terms of a system that will
offer a necessary degree of social order as well as a
different set of motives and objectives. The order that
comes to my mind as most likely to satisfy these requirements is one that blends a "religious" orientation

with a "military" discipline. Such a monastic organization of society may be repugnant to us, but I suspect it offers the greatest promise of making those enormous transformations needed to reach a new stable socio-economic basis.[3]

No part of my book has aroused more dismay that this prognosis. Here I can offer only one softening suggestion. The line between coercion and cooperation, or between necessity and freedom, is not an easy one to draw; there are armies of conscripts and armies of volunteers; churches built on dogma and churches that rest on a consensus of freely expressed beliefs. The degree of harsh authority, in other words, depends on the extent of willing self-discipline. This offers the possibility that beyond the inflection point a generation ahead we may find a variety of responses similar to those that followed the fall of the Roman Empire. Some nation-states, endowed with strong traditions of social unity, or blessed with the good fortune of political genius, may make their adaptations and transformations with a minimum of repressive force. Others may stagger from disaster to disaster, lurching from the pole of totalitarianism to that of anarchy.

It would be a disservice to the very chances of democracy to pretend that its institutions have an easy

3. These last pages of this Afterword are taken, with a few emendations, from "Second Thoughts on the Human Prospect," an essay written about a year after publication.

chance of piloting us through the protracted trials ahead. But it would be equally wrong to toss aside the possibility altogether. This is simply a question that will have to be left for the future. Our task will be to practice and strengthen the democratic way in the relatively easy years ahead. That may prove difficult enough.

There remains one last area in which "second thoughts" seem appropriate. This is the question raised at the beginning of this book, when I discussed the difficulties of putting ourselves at a sufficient distance from the long-term perspective that the human prospect forces upon us. More specifically, it concerns the attitude of stoical disengagement, of rueful but removed commentary, with which I discuss matters whose advent fills me with dismay. Is this a responsible attitude to take? Is there not a more activist philosophy that would still be appropriate to the diagnosis of the text?

I must confess that I have worried over this aspect of *The Human Prospect* more than over any of its premises or conclusions. I have been concerned lest my attitude lead to a self-fulfilling prophecy of defeat, to a cowardly passivity, to an unbearable conflict of hopes and fears. Yet after much self-search I find that I have not changed my mind on this central point, and in these last pages I must defend my position as best I can.

174

Let me begin by freely acknowledging the deep contradictions that my attitude seems to involve me in. As I examine the prospect ahead, I not only predict but I prescribe a centralization of power as the only means by which our threatened and dangerous civilization will make way for its successor. Yet I live at a time when I am profoundly suspicious of the further gathering of political power. So, too, my analysis leads me to place my hopes for the long-term survival of man on his susceptibility to appeals to national identity and to his willingness to accept authority. But my own beliefs incline me strongly in the opposite direction, detesting the claims of patriotism and mystical national unity, averse to hierarchies of sub- and super-ordination. Or take finally my judgments with respect to industrial civilization itself. Am I not the child and beneficiary of this civilization? Can I discuss its death throes unaware that I am talking about my own demise?

These and other contradictions are inextricably lodged in the human prospect as I have outlined it. Yet, curiously, I do not find myself unduly weighed down or paralyzed by these conflicts. Perhaps this is because the human prospect involves us in considerations affecting the fairly distant future, and these considerations, both for better and for worse, do not greatly influence the decisions by which we live our daily lives. If this myopia weakens our ability to prepare for the future, it also saves us the agony of seek-

ing to reconcile present behavior with future require-
ment. For in my daily life I find that I do not have
much difficulty in knowing what course to follow. As
Thornton Wilder has written, I know that every good
thing stands at the razor edge of danger and must be
fought for at every moment. Within the scope of my
daily life, I have few doubts as to what these good
things are, or what steps I must take to preserve them
from danger. (Let me add, only in passing, that I generally favor policies that would be called "democratic
socialist.")

It is only when I look to the future and ask
whether I can reconcile my daily life with the prospect
ahead that I face the moral and existential problems I
have described. For then indeed I experience the hol-
low recognition that perhaps no such reconciliation is
possible and that I may live in a time in which no
congruence can be established between the good
things of the present and the necessary things of the
future.

Such a point of view strikes us as "defeatist,"
intolerable, almost wicked. Is it? Let us take a moment
to compare the period of the disintegration of the Roman Empire with ours. Certain analogues and corre-
spondences are obvious. Then, as now, we find order
giving way to disorder; self-confidence to self-doubt;
moral certitude to moral disquiet. There are resem-
blances in the breakdowns of cumbersome economic
systems, in the intransigence of privileged minorities.

One is tempted to ask if revolutionary socialism is our Christianity; China and North Vietnam our Goths and Visigoths; the Soviet Union our Byzantium; the corporation and ministries our latifundia?

But the deeper parallel is encountered when we ask what consistent moral stance we could recommend to a person of good will who found himself or herself in the fourth century A.D. Would we urge that he or she join the Christian sect, particularly if we had foreknowledge of the toll that Christianity would exact in lives and free spirit? Would we urge the defense of the intellectual heritage of Greece or republican Rome, when it was clear beyond doubt that these ideas had run their course and no longer had the power to conjure up belief and command action? Would we recommend a futile rearguard effort to persuade a dissolute upper class to carry out long overdue reforms? And, finally, could we expect a cultivated citizen of Rome to go over to the barbarians?

To ask such questions is to confront the fact that there are periods in history in which it is not possible to reconcile the hopes of the moment and the needs of the future, when a congruence between our personal lives and the collective direction of all mankind cannot be established without doing violence either to our existence or our understanding. I believe that the present is such a time and that we must learn to live with its irreconcilable conflicts and contradictions. These conflicts and contradictions fill me with discomfort,

177

but less so than any simpler or more consistent alternative that I can construct for myself.

This is the conclusion to which my analysis of the human prospect drove me eight years ago, and it remains the conclusion to which it drives me today. I may complain at this state of affairs, but I cannot change it, just as Atlas, too, complained unendingly at the task that had been thrust upon him, but could not change that. To accept the limitation of our abilities, both as individuals and as a collectivity, seems to be the most difficult idea that Promethean man must learn. But learn it he must and learn it he will. The only question is whether the teacher will be history or ourselves.

POSTSCRIPT

What Has Posterity Ever Done for Me?

WILL MANKIND survive? Who knows? The question I want to put is more searching: Who cares? It is clear that most of us today do not care—or at least do not care enough. How many of us would be willing to give up some minor convenience—say, the use of aerosols—in the hope that this might extend the life of man on earth by a hundred years? Suppose we also knew with a high degree of certainty that humankind could not survive a thousand years unless we gave up our wasteful diet of meat, abandoned all pleasure driving, cut back on every use of energy that was not essential to the maintenance of a bare minimum. Would we care enough for posterity to pay the price of its survival?

I doubt it. A thousand years is unimaginably distant. Even a century far exceeds our powers of empathetic

---

About a year after *The Prospect* was published, the *New York Times* asked if I would do a short piece on the value of survival. I append this to the new edition of the *Inquiry* for reasons that I hope are self-evident. It needs no Afterword.

179

imagination. By the year 2075, I shall probably have been dead for three quarters of a century. My children will also likely be dead, and my grandchildren, if I have any, will be in their dotage. What does it matter to me, then, what life will be like in 2075, much less 3075? Why should I lift a finger to affect events that will have no more meaning for me seventy-five years after my death than those that happened seventy-five years before I was born?

There is no rational answer to that terrible question. No argument based on reason will lead me to care for posterity or to lift a finger in its behalf. Indeed, by every rational consideration, precisely the opposite answer is thrust upon us with irresistible force. As a Distinguished Professor of political economy at the University of London has written in *Business and Society Review:*

> Suppose that, as a result of using up all the world's resources, human life did come to an end. So what? What is so desirable about an indefinite continuation of the human species, religious convictions apart? It may well be that nearly everybody who is already here on earth would be reluctant to die, and that everybody has an instinctive fear of death. But one must not confuse this with the notion that, in any meaningful sense, generations who are yet unborn can be said to be better off if they are born than if they are not.

Thus speaks the voice of rationality. It is echoed in the book *The Economic Growth Controversy* by a Distinguished Younger Economist from the Massachusetts Institute of Technology:

. . . Geological time [has been] made comprehensible to our finite human minds by the statement that the 4.5 billion years of the earth's history [are] equivalent to once around the world in an SST. . . . Man got on eight miles before the end, and industrial man got on six feet before the end. . . . Today we are having a debate about the extent to which man ought to maximize the length of time that he is on the airplane.

According to what the scientists now think, the sun is gradually expanding and 12 billion years from now the earth will be swallowed up by the sun. This means that our airplane has time to go round three more times. Do we want man to be on it for all three times around the world? Are we interested in man being on for another eight miles? Are we interested in man being on for another six feet? Or are we only interested in man for a fraction of a millimeter—our lifetimes?

That led me to think: Do I care what happens a thousand years from now? . . . Do I care when man gets off the airplane? I think I basically [have come] to the conclusion that I don't care whether man is on the airplane for another eight feet, or if man is on the airplane another three times around the world.

Is this an outrageous position? I must confess it outrages me. But this is not because the economists' arguments are "wrong"—indeed, within their rational framework they are indisputably right. It is because their position reveals the limitations—worse, the suicidal dangers—of what we call "rational argument" when we confront questions that can only be decided by an appeal to an entirely different faculty from that of cool reason. More than that, I suspect that if there is cause to fear for man's survival it is because the calculus of logic and reason will be applied to problems where they have as little validity, even as little bearing, as the calculus of feeling or sentiment applied to the solution of a problem in Euclidean geometry.

If reason cannot give us a compelling argument to care for posterity—and to care desperately and totally—what can? For an answer, I turn to another distinguished economist whose fame originated in his profound examination of moral conduct. In 1759, Adam Smith published "The Theory of Moral Sentiments," in which he posed a question very much like ours, but to which he gave an answer very different from that of his latter-day descendants.

Suppose, asked Smith, that "a man of humanity" in Europe were to learn of a fearful earthquake in China—an earthquake that swallowed up its millions of inhabitants. How would that man react? He would, Smith mused, "make many melancholy reflections upon the precariousness of human life, and the vanity of all the labors of man, which could thus be annihilated in a moment. He would, too, perhaps, if he was a man of speculation, enter into many reasonings concerning the effects which this disaster might produce upon the commerce of Europe, and the trade and business of the world in general." Yet, when this fine philosophizing was over, would our "man of humanity" care much about the catastrophe in distant China? He would not. As Smith tells us, he would "pursue his business or his pleasure, take his repose or his diversion, with the same ease and tranquillity as if nothing had happened."

But now suppose, Smith says, that our man were told he was to lose his little finger on the morrow. A very different reaction would attend the contemplation of this "frivolous disaster." Our man of humanity would be reduced to a tormented state, tossing all night with fear and dread—whereas "provided he never saw them, he will snore with the most profound security over the ruin of a hundred millions of his brethren."

Next, Smith puts the critical question: Since the hurt to his finger bulks so large and the catastrophe in China so small, does this mean that a man of humanity, given the choice, would prefer the extinction of a hundred million Chinese in order to save his little finger? Smith is unequivocal in his answer. "Human nature startles at the thought," he cries, "and the world in its greatest depravity and corruption never produced such a villain as would be capable of entertaining it."

But what stays our hand? Since we are all such creatures of self-interest (and is not Smith the very patron saint of the motive of self-interest?), what moves us to give precedence to the rights of humanity over those of our own immediate well-being? The answer, says Smith, is the presence within us all of a "man within the breast," an inner creature of conscience whose insistent voice brooks no disobedience: "It is the love of what is honorable and noble, of the grandeur and dig-

183

nity, and superiority of our own characters.''

It does not matter whether Smith's eighteenth-century view of human nature in general or morality in particular appeals to the modern temper. What matters is that he has put the question that tests us to the quick. For it is one thing to appraise matters of life and death by the principles of rational self-interest and quite another *to take responsibility for our choice.* I cannot imagine the Distinguished Professor from the University of London personally consigning humanity to oblivion with the same equanimity with which he writes off its demise. I am certain that if the Distinguished Younger Economist from M.I.T. were made responsible for determining the precise length of stay of humanity on the SST, he would agonize over the problem and end up by exacting every last possible inch for mankind's journey.

Of course, there are moral dilemmas to be faced even if one takes one's stand on the ''survivalist'' principle. Mankind cannot expect to continue on earth indefinitely if we do not curb population growth, thereby consigning billions or tens of billions to the oblivion of nonbirth. Yet, in this case, we sacrifice some portion of life-to-come in order that life itself may be preserved. This essential commitment to life's continuance gives us the moral authority to take measures, perhaps very harsh measures, whose justification cannot be found in the precepts of rationality, but must be sought in the unbearable anguish we feel if we imagine ourselves as the executioners of mankind.

This anguish may well be those "religious convictions," to use the phrase our London economist so casually tosses away. Perhaps to our secular cast of mind, the anguish can be more easily accepted as the furious power of the biogenetic force we see expressed in every living organism. Whatever its source, when we ask if mankind "should" survive, it is only here that we can find a rationale that gives us the affirmation we seek.

This is not to say we will discover a religious affirmation naturally welling up within us as we career toward Armageddon. We know very little about how to convince men by recourse to reason and nothing about how to convert them to religion. A hundred faiths contend for believers today, a few perhaps capable of generating that sense of caring for human salvation on earth. But, in truth, we do not know if "religion" will win out. An appreciation of the magnitude of the sacrifices required to perpetuate life may well tempt us to opt for "rationality"—to enjoy life while it is still to be enjoyed on relatively easy terms, to write mankind a shorter ticket on the SST so that some of us may enjoy the next millimeter of the trip in first-class seats.

Yet I am hopeful that in the end a survivalist ethic will come to the fore—not from the reading of a few books or the passing twinge of a pious lecture, but from an experience that will bring home to us, as Adam Smith brought home to his "man of humanity," the personal responsibility that defies all the homicidal promptings of reasonable calculation. Moreover, I believe that the coming generations, in their encounters with famine, war, and the threatened life-carrying capacity of the globe, may be given just such an experience. It is a glimpse into the void of a universe without man. I must rest my ultimate faith on the discovery by these future generations, as the ax of the executioner passes into their hands, of the transcendent importance of posterity for them.

Acknowledgments

I AM INDEBTED TO many people for their help in writing this book—some for technical advice, some for detailed criticism, some simply for their counsel and encouragement. I will not embarrass my friends and colleagues by revealing the extent of their complicity in this book. Let me therefore list them alphabetically: Moses Abramovitz, Daniel Bell, Stanley Burnshaw, Paul Ehrlich, John Holdren, Arien and Irving Howe, Robert Silvers, Hans Staudinger, Thomas Vietorisz. One name only I place out of order because I know that its bearer is willing to be excused from the usual disclaimer that frees all the above from any responsibility for the content of these pages. As always, I salute Adolph Lowe, the spiritual co-author of these pages and the original source of the metaphor of Atlas.[1] Finally, let me thank my secretary, Ms. Salzman. I do not think there is a Perfect Secretary in the Greek gallery of the gods, but if there were, her name would surely be Lillian.

1. See "S ist noch nicht P," in *Ernst Bloch zu ehren* (Frankfurt am Main, 1965), p. 142. Professor Daniel Bell has pointed out that Bertrand Russell (unbeknownst to Professor Lowe or myself) used the metaphor of "a weary but unyielding Atlas" in a moving conclusion to his essay "A Free Man's Worship" (1903).

A brief word on sources. I have sought to identify all quotations and to underpin with citations those parts of the argument in which scientific expertise is critical. I have not burdened the text with appeals to authority for the economic, sociological, or psychoanalytic sections of the text, save in occasional places where a reader might wish to know of a parallel argument. So much of the book is built on the foundation of my life as well as my studies that extensive footnoting seemed pointless: unnecessary to those who will accept the message of the book as it now stands, unavailing to those who will not.

Index

Acton, Lord, 145
Aristotle, 145
Atlas, 165, 166
attitudinal changes, 14–16, 26–27
authoritarianism, 39, 105–7, 128–29, 132–33
authority, attitudes to, 122, 125–29, 132–33, 136–37, 153–54
Ayres, Robert, 51, 51n., 54n.

Bethe, Hans, 40, 41n.
biosphere, 73–74, 170
birth control, 33, 38–39, 62
blackmail, nuclear, 43–46, 95
Bogue, David, 62n.
"Blueprint for Survival," 155n.

Calleo, David, 121, 121n.
capitalism, 79ff.
    authoritarianism and, 105–7
    crisis in, 111
    ethos of, 18, 19, 101f.
    external challenges to, 94–107
    and democracy, 105–7
    and growth, 98f.
    ideal type, 79, 80, 82f., 106
    and income distribution, 101–3
    instability of, 113, 114–16
    scenarios, 82, 84–85
    services in, 103–4
    stationary state, 99f.
    and underdeveloped world, 95–96, 97
China, 33, 34, 42, 91, 127–8, 132, 142
civilizational malaise, 18–19, 27, 163 (*see also* industrial civilization)

climatic change, 51–53
Cloud, Preston, 49n., 50n.
Club of Rome, 155
coal, 69
Commoner, Barry, 37n.
conservative view of man, 137f.
convulsive change, 154

Daly, Herman, 49n.
Davis, Kingsley, 33n.
democracy, 105–7, 108, 128–29
demographic problems, 31–36
    and capitalism, 95–96
    and socialism, 96, 97
Diggins, John, 162n.
Dinnerstein, Dorothy, 148n.
distribution of income, 101f.
domination, 145–46
doomsday, 160

Easterlin, Richard, 86n.
Eberstadt, Nick, 62n.
Edwards, R. C., 143n.
Ehrlich, Paul, 36, 36n., 49n.
Eisenberg, Leon, 138, 138n.
energy crisis, 17, 50–56, 68–71
Engels, Friedrich, 147
environmental challenge, 47ff., 68, 73–74, 170
equality, 127–28

Falk, Richard, 44, 44n.
famine, 63
fluorocarbon, 73
Flower, Andrew, 69n., 168n.
Frejka, Tomas, 34n.
Freud, Sigmund, 124, 136
Frisken, W. R., 51n., 52n.

Index

Garaudy, Roger, 80, 80n.
Great Crash, 28
Great Depression, 116
greenhouse effect, 72
Green Revolution, 36
growth, limits of, 17, 46f., 98f., 110, 155–56

heat, 50–56
Hobbes, 145
Hobbesian struggle, 105–6, 108
Holdren, John, 49n., 51n.
human nature, 123, 124–25, 130–31, 136–43, 153

ideal types, 80f., 106, 110
identification, 122, 130–31, 133–36
income distribution, 101f.
industrial civilization, 91–94, 108–11, 160–63, 175 (*see also* capitalism, socialism)
inflation, 28, 112–18
intellectual "sentries," 159–60
Issawi, Charles 69n., 168n.

Keynes, John Maynard, 84, 100
Kneese, Allen, 51, 51n., 54n.
Komanoff, Charles, 70n.
Kovel, Joel, 148n.

La Rochefoucault, 145
Lasswell, Harold, 123n.
Leon, Paolo, 86n.
Leontief, Wassily, 168–170
Limits to Growth, 155
Lovering, T. S., 49n., 50, 50n.

Machiavelli, 145
Mackintosh, N. J., 145n.
Marshall, Alfred, 84
Marx, Karl, 100, 113, 141, 147, 161
Marxian view of capitalism, 82–84
military socialism, 39
mood, causes of present, 12–16, 18–19, 25–28

national identification, 122, 130–36, 142

nation-states, 46, 121, 131f.
negative population growth, 60
nuclear war, 40–41, 46, 65, 95
nuclear weaponry, 40–43, 65, 75

obedience, 122, 125–29, 130–33, 136–37
oil, 66–67, 68–70, 111, 112–13, 167–68
Ollman, Bertell, 141–142, 142n.
Organization of Petroleum Exporting Countries (OPEC), 66–67, 70, 71, 111, 112, 114, 167

peoplehood, 131–36
Pilbeam, David, 145n.
Plato, 145
Platt, John, 154, 154n.
political element in man, 122–23, 125–29, 130f.
political power, 120–121, 144–148, 175
pollution, 50–56
population, 31–36, 59–64, 75, 95–97, 184,
posterity, 179–186
post-industrial society, 110, 142, 160–63
primitive society, 163–64
Promethean spirit, 164

radical view of man, 137–44
rationality and future concern, 134–35
resources, 47–50, 70, 167
revolutionary governments, 39, 42–46, 127
Richardson, L. F., 45n.
Rowland, Benjamin, 121, 121n.

science, 56–58, 92–94, 108–10
services, 103–4
Smith, Adam, 182–84, 185
social harmony, 85–86, 90
socialism, 91–95, 107–111, 117
  and democracy, 108
  ideal types, 86–89, 89f.
  and technology, 110–11

in underdeveloped world, 39, 127–29, 132–33
socio-economic analysis, 79–81, 119
stationary state, 99f.
stock market, 28
subjective elements in analysis, 20–23
Sweezy, Paul M., 80, 80n.

Talleyrand, 23
technology, 56–58, 92–94, 108–10
terrorism, nuclear, 43, 45
thermal pollution, 50–56
Tsu, Amy Ong, 62n.

underdeveloped world, 31–36, 37–38, 42–46, 61–62, 94–98, 169

United States as "pure" capitalism, 79–82
USSR and socialism, 80–81, 87–88, 90

value-free inquiry, 20–23

wage and price controls, 117
wars, 40–46, 97, 157
of "preemptive seizure," 43, 97
of redistribution, 43–46, 95
Wilde, Oscar, 26
Wilder, Thornton, 176
Willrich, Mason, 42n.
Wilson, E. O., 140–41, 141n., 145n.

zero population growth, 33, 34, 59–61, 64